TECHNOLOGY AND ECONOMIC POLICY

TECHNOLOGY AND ECONOMIC POLICY

edited by
RALPH LANDAU
DALE W. JORGENSON

BALLINGER PUBLISHING COMPANY
Cambridge, Massachusetts
A Subsidiary of Harper & Row, Publishers, Inc.

International Standard Book Number: 0-88730-068-5

Library of Congress Catalog Card Number: 86-14025

Printed in the United States of America

Library of Congress Cataloging-in-Publication Data

Technology and economic policy.

Includes bibliographies and index.
1. Technological innovations—Economic aspects—United States—Congresses. 2. Technology and state—United States—Congresses. 3. Taxation—United States—Congresses. I. Landau, Ralph. II. Jorgenson, Dale Weldeau, 1933-
HC110.T4T38 1986 338'.06 86-14025
ISBN 0-88730-068-5

CONTENTS

LIST OF FIGURES

LIST OF TABLES

PREFACE

This volume is the first publication from the Program on Technology and Economic Policy, established at the John Fitzgerald Kennedy School of Government, Harvard University, in September 1984. This program is intended to fill an important gap in the debate over economic policy by providing a technological perspective. The primary focus of the program is on the contribution of technology to economic growth and the impact of public policies on technology.

The ultimate goal of the Program on Technology and Economic Policy is to improve the formulation and implementation of policies to stimulate technological innovation. The program's initial objective is to bring existing research on technology to bear on current issues in economic policy. Although a great deal of worthwhile research is already available, discussions of economic policy have failed to take advantage of it.

The present volume is based on the proceedings of a Conference on Technology and Economic Policy, held in Washington, D.C., on February 4–5, 1985. The purpose of the conference was to develop an understanding of the interrelationships between current issues in economic policy and technological innovation. In view of the high priority assigned to tax policy by the Reagan administration, more than half of the conference was devoted to business taxation.

The conference considered both the overall impact of business taxation and the impact of specific tax provisions on the rate of innovation. The consideration of business taxation focused on the role of the corporation income tax. Specific tax provisions that were considered included those affecting the supply of venture capital, the taxation of capital gains, and the recovery of capital costs.

The conference devoted considerable attention to the large and growing deficits of the federal government. The high level of the deficit has had an important impact on the financing of investments in new technology. The Congress has now passed legislation to try to bring the deficit under control by arresting the growth of government spending.

The U.S. economy is faced with switching from a creditor to a debtor in world financial markets. Policies to restore the international competitiveness of the U.S. economy were much discussed at the conference. Special emphasis was given to the crucial role of the cost of capital in the United States and its major trading partners in determining international competitiveness.

To sustain the initial momentum resulting from the February 1985 conference, the Program on Technology and Economic Policy sponsored a second conference in November 1985. This conference dealt with technology and capital formation. The conference reviewed current research on this topic and developed an agenda for future research oriented toward the needs of policymakers.

1 TECHNOLOGY, ECONOMICS, AND PUBLIC POLICY

Ralph Landau[1]

TECHNOLOGICAL INNOVATION IN U.S. ECONOMIC HISTORY

Understanding the interactive nature of technology and economics can lead to a better insight into public policy priorities. The modern economics profession awoke to a comprehension of the major role technology has played only in the middle 1950s, when two economists, Robert Solow of the Massachusetts Institute of Technology and Moses Abramovitz of Stanford University, published works examining the sources of American economic growth since the Civil War. By comparing the conventional inputs with the outputs of the economy for the century following that conflict, they found the inputs were far smaller than the outputs, the difference being attributed to a residual they equated with "technological progress," which seemed to constitute nearly *90 percent* of the growth during the period from the Civil War until the midfifties. Other papers followed, by D. Jorgenson of Harvard University, J. Kendrick of George Washington University, N. Rosenberg and K. Arrow of Stanford University, E. Mansfield of the University of Pennsylvania, R. Nelson of Yale, B. Klein of the California Institute of Technology, E. Denison of the Brookings Institution, and others, in the sixties and seventies. It is now widely believed that from *one-third to one-half of all our growth has come from technological progress,* and that it is the principal driving force for long-term economic growth and the increased standards of living of modern industrial societies.[2] Indeed, periods of rapid technological growth tend to be correlated with periods of rapid economic growth. Moreover, technology enables exporting and competition in the international marketplace, improves productivity, and is vital for the defense of the country.

Thus, technological innovation and change create jobs and investment opportunities and enhance human productivity, thereby raising the standard of living and the wealth of society as a whole. There is little disagreement among economists, politicians, and technologists that this is so, regardless of the rise and fall of business generally in business cycles, but the rate of exploitation of technology is extremely sensitive to national economic and social policies. In an era of major new technological developments, it is especially important to analyze such sensitivities and required policies.

History has shown that economic growth results both from slow, steady improvements in technology and knowledge (embodied in physical and human capital), as well as from "breakthrough" innovations. Each type is probably of equal importance. However, because breakthrough innovations are unpredictable, so are their scope, dimensions, and economic effects. Such innovations may change directions of entire industries, or create new industries. Some generate new jobs and change the growth rate of the economy significantly, expanding markets to new heights. The steam engine and the railroad created low-cost mass markets; the automobile fostered spending on housing, highways, energy, and appliances, and it created jobs. The jet airplane has stimulated the movement and mobility of people and the world tourist industries, distributing wealth worldwide. The transistor and the computer are two modern innovations whose impact on the economy of the future will be even greater than was the impact of the electrification of America in the twentieth century. Technologically based innovations are likewise leading to a surge in employment in such diverse areas as construction, supplies, and servicing of office buildings. Even fields like apparel, preprocessed foods, and residential and day care services, have been stimulated, because the newer technologies offer more job opportunities for women. There has been a great increase in job training and education at many levels. Financial services have been revolutionized; insurance, banking, brokerage, and many other institutions have grown and increased their productivity and innovation although not yet at the level of manufacturing and agriculture.

Table 1–1 lists some of the important breakthroughs, not all of equal economic significance. Under pharmaceuticals, recombinant DNA technology clearly builds on one of the great scientific advances of the century, as did quantum physics in the twenties and thirties. The latter was the basis of lasers, radar, television, fiber optics, semiconductors, and many electronic marvels. The full industrial implications of recombinant DNA technology are probably twenty years off.

Throughout history, economists and others have been predicting imminent large-scale technologically induced unemployment which never materialized, although some jobs (like blacksmithing) disappeared, only to be replaced by others as the wealth of the economy, produced substantially by technological change, increased and new industries and companies appeared—created by the unique *entrepreneurial American spirit, in a positive-sum game* (Ralph Landau and Nathan Rosenberg 1986). Similarly, today, new jobs for new entrants into the labor force are being created faster than old ones disappearing from the smokestack industries. In fact, even many of those recently displaced have already been employed elsewhere. The U.S. Labor Department has shown that, of 5.1 million workers who lost jobs in heavy industry from 1979 through 1983, 3.1 million had jobs in January 1984, and 700,000 had retired. Not all the new jobs were as good as the old ones, but they *were* working. The extraordinary success of the United States in creating jobs for a rapidly expanding work force is illustrated in Table 1–2, comparing postwar Europe recovering from war devastation with the effective Japanese efforts.

The power of compounding is great; with a growth rate of only 2 percent per year in real income per person, standards of living can almost double between generations. This was accomplished in the United States in the century after the Civil War by an average productivity increase of 1.8 percent per year, while population grew enormously; and the United States moved from a largely rural economy to the greatest industrial power in the world. In the process, it overhauled the then champion, the United Kingdom, which grew at only 1 percent per year, with a productivity increase averaging 1.2 percent per year. In 1864, the British pound sterling was at its all time high, worth $9.92 in American currency. It has been downhill ever since; it nearly approached $1.00 recently! Japan's experience is much more like that of the

Table 1–1. Some Breakthrough Innovations since 1945

Pharmaceuticals	**Communications**
Antibiotics	Optical fibers
Hypertension drugs	Satellite systems
Anti-ulcer drugs	Television
Birth control pills	Stereo sound
Vaccines	**Transportation**
Recombinant DNA technology	Jet airplanes
Chemicals	Space vehicles
Herbicides	Lead-free gasoline
Synthetic fibers	**Reproduction**
Plastic films for packaging	Xerography
Electronic Devices	Instant photography
Transistors	**Materials**
Integrated circuits	Plastics
Lasers	Alloys, ceramics, etc.
Solar cells	Composites
Videotapes and disks	**Household Appliances**
Magnetic recordings	Air conditioners
Microchips	Electric clothes washers and dishwashers
Computers	**Factory Production**
Mainframe large computers	Tape-controlled machine tools
Minicomputers	Robots
Microprocessors	Powder metal technology
Hand calculators	
Software systems	

United States.

During this lengthy period, government played a relatively small role and spent only a few percent of the gross national product, although it created conditions that were helpful to growth, such as the land grant institutions and railroads of the nineteenth century and the antitrust actions and educational and basic research innovations of the twentieth.

It was the entrepreneurial efforts of private citizens and the continuous appearance of new companies that produced this historic change. By encouraging the evolution of many different types of business organizations, large and small, the United States found an effective solution

Table 1–2. Employment, Civilian Millions

	EEC[a]	United States	Japan
1955	101.4 Est.	62.2	41.9
1965	104.8 Est.	71.1	47.3
1975	102.2	85.8	52.2
1984	101.6	104.2	58.0
(Projected gain)	0	42	16

a. The Ten

Source: OECD, EEC, Bank of Japan, IFO (from a paper by Rupert Pennant-Rea, *The Economist*).

to rapidly changing environments. The system could adapt more efficiently than the old and rigid structures of Europe, and thus seemed to follow the Darwinian principles of evolutionary adaptation which permitted animal and human life as we know it to survive and prosper, while the dinosaurs became extinct. Today we see this flexibility in the shift from smokestack mass production industries run by professional managers to the new era of technological innovation and entrepreneurship embodying brand new methods in the workplace and the office.

UNDERSTANDING THE INNOVATION PROCESS

Inadequate appreciation of the close tie between technology, economics, and growth and the entrepreneurial climate in which they flourish seems to have contributed to the inability to understand how the dilemmas of the 1970s—raging inflation, low growth, high interest rates, productivity decrease, elevated unemployment—could be overcome and the more traditional patterns resumed. This intellectual lag is illustrated by the difficulties encountered in the new companies' activities beginning in the late 1960s. It culminated in 1978 in the Carter administration's proposal to eliminate the few vestiges of favorable capital gains tax treatment, still in pursuit of the political egalitarianism which had dominated the socioeconomic philosophy of the 1960s. This brought about a major national debate. Entrepreneurs, previously so effective in generating new technology and jobs, recognized that the steady rise in the capital gains tax was adversely affecting various sources of private capital, including the venture capital industry, which is unique to the United States and is a major factor in the creation and nurturing of new companies, with attendant technological and other risk-taking. In 1969 the capital gains marginal tax had been raised from 25 percent to 35+ percent, and in 1976 from 35+ percent to a maximum of 49.1 percent. As a result, the amount of new venture capital commitments in the United States dropped to as low as $10 million in the year 1975, from $171 million in 1969. The number of new public issues of young companies almost disappeared, diminishing from about 1,000 in 1969 to as few as twenty or so per year during the 1973–1977 period.

If emphasis were needed that science, technology, and engineering are not enough to create commercially successful innovations, these statistics seemed to make the case. Innovation consists of two stages: invention and implementation. Usually, the latter is the riskier and the more expensive. Capital is nearly always required to put good ideas into practice. During the 1970s, young companies starved for capital were forced in many cases to sell all or part of themselves to Japanese and other foreign investors, or to license valuable technology abroad. These foreign companies lost no time in further implementation of the inventions so accessible to them. It has become stark fact that the Japanese are very good at implementation indeed. Hence, the response from those dependent on venture capital in Silicon Valley and along Route 128 and other entrepreneurial industries was prompt and surprisingly effective. The outcome was that legislation was adopted in 1978 to reduce the capital gains tax from 49.1 percent to 28 percent.

Data released by the U.S. Treasury in an extensive September 1985 report on the effects of the 1978 capital gains tax reduction (Figure 1–1) show the history of actual capital gains taxes paid to the government since 1962. From a high of nearly $6 billion in 1968, the receipts from this source fell to just over $4 billion by 1975, then began to rise steadily, approaching $13 billion by 1982. Thus, reducing the capital gains tax in 1978 actually produced nearly $4 billion more tax revenue, contrary to dire Treasury predictions of a $2 billion loss.

The Treasury report (1985:115) says that ". . . the capital gains tax reductions of 1978 appear to have increased long run realizations enough to offset the revenue losses from lower capital gains tax rates." Conversely, it also says, "any effort to raise substantial revenue by increasing capital gains tax rates, given the current parameters of the tax laws, is likely to be ineffective." While there was certainly an unlocking effect as old investments were cashed in under this more

Figure 1–1. Year Taxes Paid on Capital Gain.

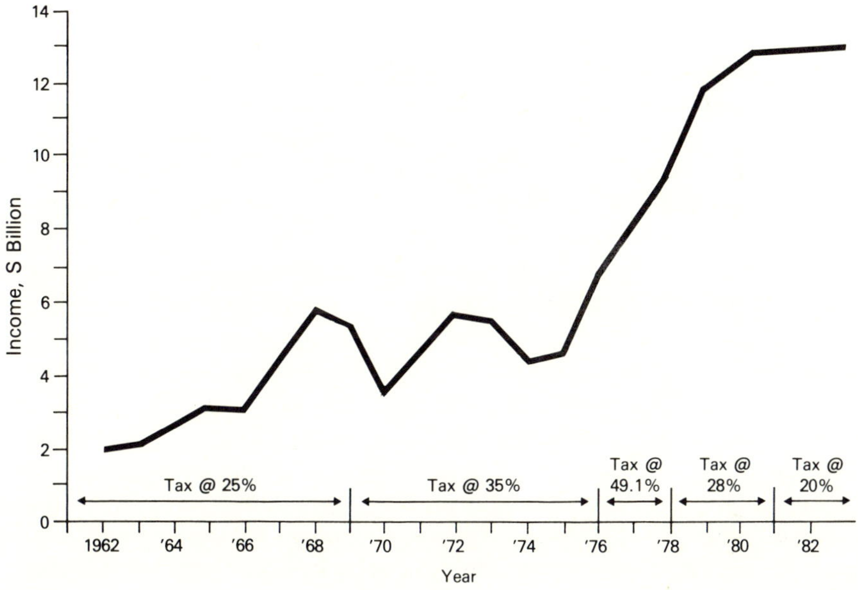

favorable tax situation, obviously the stimulation for risk-taking and investment in new enterprises was also significant; the dynamic nature of the economy was once more demonstrated. Invention and implementation could proceed at an accelerating pace. The amount of new venture capital raised by independent private venture capital firms increased sharply to an annual rate of almost $3.4 billion per year in 1983 and about $3.2 billion in 1984. When Small Business Investment Companies (SBIC) and corporate funds are included, new capital commitments to the venture capital industry were $4.2 billion in 1984. The new issue market zoomed ahead of Wall Street as a whole; smaller company new issues totaled $5.7 billion in 1983, the highest ever, but declined to $1.7 billion in 1984, when the new issues stocks fell out of favor. The situation is improving in 1985. Total new capital through initial public stock offerings rose from only $153 million in 1977 to

$12.6 billion in 1983. Even the reduced 1984 offerings were higher than any years preceding 1983. Total funds raised by corporations through common stock offerings rose from $6 billion to $32 billion in the same period. Research and development partnerships arose, encouraged by the U.S. Commerce Department, and generated another $1 billion or so by 1983 (1984 saw a decline as the risks became greater). New high-technology and a much larger number of low- and medium-technology firms either appeared, or started to grow rapidly, in spite of inflation's bad effect on capital formation generally. Figures 1—2 and 1—3 illustrate these trends. Equity values as a whole rose 24 percent from January 1978 to December of 1979, and 33 percent more in 1980. After inflation adjustment, the total rise was about 30 percent. Encouraged by this surge of private investment in risk-taking equities and enterprises, and by favorable rulings from the Department of Labor, pension funds and other institutions, foreign and domestic, allocated a modest portion of their assets to such endeavors, thereby swelling the pool of available funds. However, the professional managers of these funds are very conscious that they are investing *other people's money* in risk-taking enterprises, with consequent fear of litigation in the event of failure, so that they not only limit their participation to small percentages of their portfolios, they also follow the lead of the individuals who are betting *their own money! Thus, an unprecedented large-scale economic experiment had been successfully, if unwittingly, conducted upon the entire country.* The Treasury report of September 1985 contains a very extensive analysis of many of the points raised in this chapter, and is an excellent confirmation of it, particularly as to the beneficial effects on risk-taking and investment.

MACROECONOMICS AND MICROECONOMICS

What was the true significance of this recent history? It demonstrated that technological progress, innovation, and job formation function successfully only within a larger social environment that provides an effective combination of *incentives* and complementary inputs. That is, it showed conclusively the sensitivity and responsiveness of technological innovation and change to national policy. The 1978 and 1981 tax acts demonstrate also that there was little dispute, whether politically partisan or theoretical, as to the need for policies which encouraged entrepreneurship, risk-taking investment, and technological change. This is opposite to the concept of "neutrality" (see below), which

Figure 1–2. Companies Going Public.

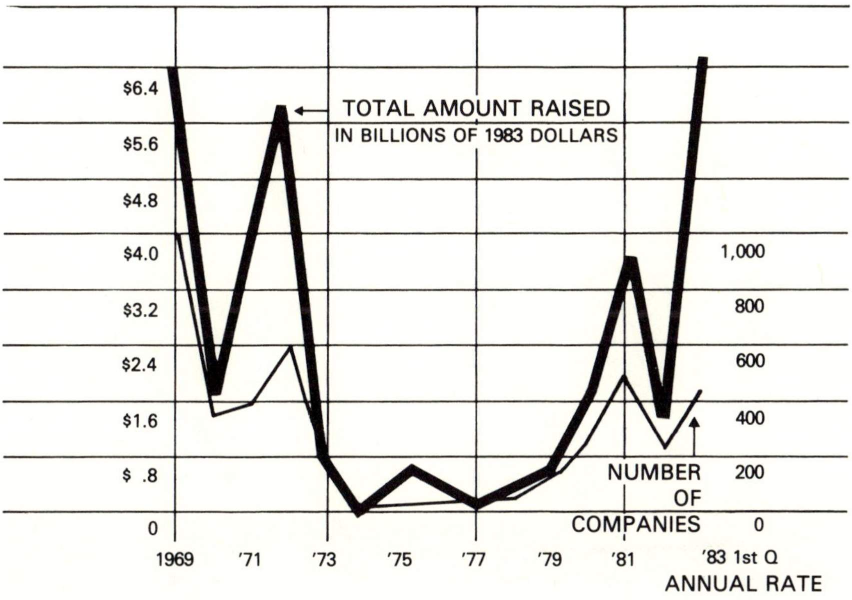

Source: "Going Public: The IPO Reporter," *Fortune,* (May 30, 1983).

makes no attempt at such optimization.

There was also significance in these events for economic theory. The models that economists have built over the postwar years to simulate the economy have been primarily devoted to studying business cycles, and, depending on the underlying theory, to developing policy for ameliorating such cycles, particularly unemployment, essentially by managing aggregate demand. The growth of the economy caused by technological change cannot be captured by these models, although there is a substantial body of economic literature which ties technological change to specific policies. That is why actual, indisputable experience such as cited in the preceding paragraphs is clearly more important and confirmatory of these theories in defining the proper climate for innovation. Nevertheless, there were more general theoretical

Figure 1–3. New Private Capital Committed to Venture Capital Firms.

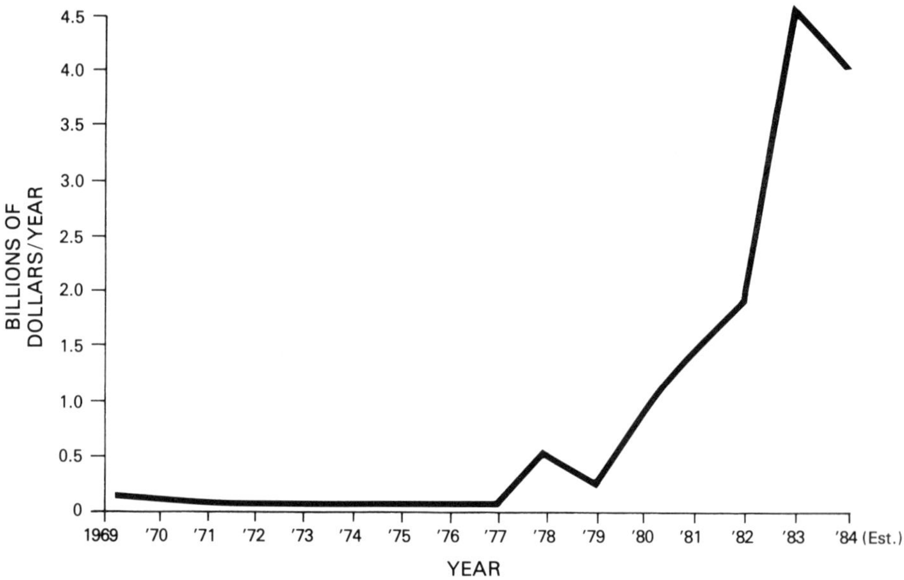

Source: Venture Economics, Inc., Wellesley Hills, Mass.

implications as well in this story. It gave the first real support to the possibility that the prevailing postwar neoclassical neo-Keynesian macroeconomics was no longer sufficient to optimize the microeconomy and that alternative microeconomic policies specifically adapted for that purpose could be found. Thus, while macroeconomists might agree on a certain fiscal policy as prudent for the government, including an appropriate proportion of taxation, microeconomists could now say that it made a very real difference to the growth of the economy whether such taxes were collected by a truly egalitarian heavily progressive system or by a selective favoritism for taxation and other policies that favored longer term dynamic economic growth, paying less attention to static fluctuations in year-to-year results. (The Kennedy round of 1964 tax cuts, more generally distributed, and of much smaller scope than the more recent cuts, had not had the same theoretical impact, partially because it was obscured by the Vietnam War and the growing inflation. Economists today still differ as to their effects, but there is general agreement these cuts were primarily intended as demand stimuli.)

Postwar government policies were not lacking microeconomic components, but they were primarily directed (especially after 1960) at social goals: alleviation of the problems of unemployment, illness, the handicapped, the aged, and eventually the poor by a redistribution of income. Those services which governments are truly required to furnish, such as defense, infrastructure, and education, suffered as a consequence, although government spending as a percentage of GNP rose. Defense spending as a percentage of total national budget dropped from 44 percent in 1969 to 24 percent in 1977; as a percentage of GNP, it fell from 8.7 percent to 5.2 percent. It was only in this way that governments could cope at all with the increasing social demands made on the economy, including control over abuses of the private sector. It had been assumed that the producing private sector could essentially take care of itself and could even withstand heavier tax and regulatory burdens, so overflowing was the American cornucopia perceived to be by politicians. (Have these notions rearisen today?) From the Second World War until the 1960s, the United States uniquely was able to realize the scientific and technological progress that had been pioneered just before, during, and immediately after the war. While R&D was also increasing, it was not yet the period of high technology, although the foundations were being laid (for example, the mainframe computer was being developed). There was a great expansion of American capacity, much of it based on such earlier developments, and its very modernity made American industry highly competitive. Of course, some sectors of the economy, like farmers, unions, and certain businesses, always managed to extract favorable treatment because of their political clout; but there was no systematic governmental intervention in the producing microeconomy.

This optimistic conception, confirmed by experience, was aided by an economic theory which focused only on aggregate demand management via the government budget balance. Under the traditional classical equilibrium analysis of optimal resource allocation with which it was combined, it treated firms and individuals as *passive agents,* while assuming thereby perfect competition so as to prevent inflation and assure the fullest possible employment. Under these postulates there would always be enough supply of human and material resources for growth (supply would follow demand), and no specific measures to assure that this was so were needed. The microeconomic world had no feedback effect on the macroeconomic world under these theories, and in this sense the dominant macroeconomic philosophy neither had nor required microfoundations. The fact that the United States found ready markets in war-exhausted Europe and the Far East, as well as at home, contributed to the mildness of the business cycles, to the modest fluctuations in the level of unemployment, to low interest rates (which encouraged risk-taking and entrepreneurship even by large firms, although the automobile industry seemed to emphasize mostly styling changes), and to reasonable wage demands by labor, which still remembered the Depression. (Their memories were symbolized by the no-strike but generous wage agreements initiated in the auto industry.)

No provision for technological change or for entrepreneurship was therefore embodied in the theory—they just happened! Economists called technology an *exogenous* variable.

The problems arose when the economy entered the period of stagflation, escalating energy costs, and suddenly declining rates of productivity increase in the 1970s. There was no longer available a stable or predictable tradeoff of unemployment versus inflation at politically acceptable levels (the Phillips Curve); market equilibrium seemed not to be attained, even in the longer run (Figure 1–4).

The rising interest rates made risk-taking and real R&D spending increases much less prevalent as financial horizons became very short-term; the equities markets languished uneasily (the real value of equities in the eighties is much less than in the sixties). But the inflation masked the growing international uncompetitiveness of many industries, as they raised prices fast enough to maintain domestic profits. Naturally, large wage increases followed. Government regulations mounted dramatically. Conflicting schools of economics consequently arose. Some revisionist neo-Keynesians felt that pure macroeconomic adjustments ("fine-tuning") could not overcome the various "stickinesses" or unpredictabilities, as they saw them, such as those of wages and prices, which prevented market-clearing equilibrium from being achieved. The neoclassical neo-Keynesians had believed that disequilibria such as unemployment would only be temporary and could be alleviated by government macroeconomic policy. Thus, without knowing what the outcome would be, certain new or revisionist Keynesians advocated direct government intervention in the microeconomy, like Nixon's unsuccessful price and wage controls. Some economists even began attempting to formulate the tenets of such Keynesian "disequilibrium" economics. Out of these efforts came proposals for short-term

Figure 1–4. Unemployment Versus Inflation (or "Where Has the Phillips Curve Gone?").

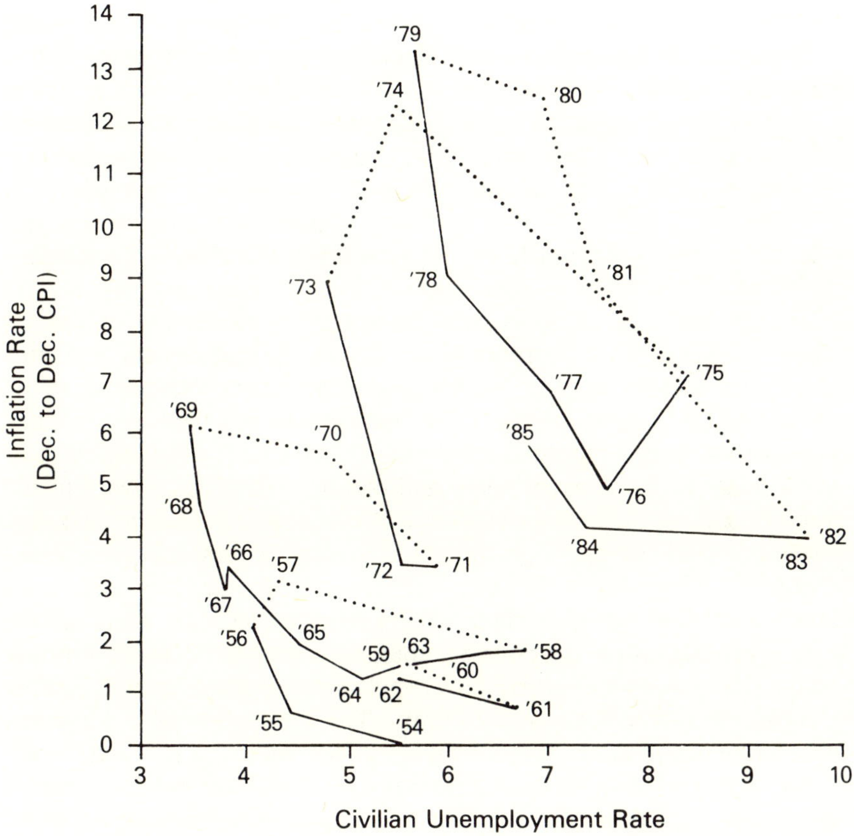

Source: Professor John Shoven, Department of Economics, Stanford University.

policy actions by government to alleviate short-term cyclical effects, such as monetary manipulation.

On the other hand, the Friedman monetarists concluded that neo-Keynesian macroeconomic tinkering was unworkable, that inflation was always and everywhere a monetary phenomenon, that there is a "natural" rate of unemployment, and that macroeconomic demand should be managed by a low predictable regular increase in the money supply, without microeconomic intervention or fine-tuning of so complex an economy. Some economists think that this can only be effective if practiced on an international scale, as the world is now in essence on a dollar standard, but political difficulties have made formal conformity impractical thus far. In practice, however, the principal international powers have acted more or less in parallel to bring inflation of the world money supply down and would no doubt act in the reverse way if reflation were to become desirable. Actual postwar monetary policy, on the contrary, had been largely accommodative to government fiscal policy. Nevertheless, the bitter experience of investors during the last fifteen years of rising inflation gave growing credibility to the monetary approach and its relation to the expected budgetary deficits, even though no precise understanding of just what "money" is has ever crystallized. The financial community's continual innovations as well as international currency flows always seemed to stay ahead of economists' definitions and predictions of M1, etc. Recent work by R.W. Hafer (1983) and others of the Federal Reserve Bank of St. Louis has explained some of these discrepancies.

As a result of these experiences and controversies, there arose in the seventies and eighties a newer school of economics called *new classical economics,* which reverted to equilibrium analysis, but also introduced the concept of *rational expectations.* In the last decade, this school has gained a significant intellectual dominance, especially among the younger generation of academic economists. Although there is a strong relation between monetarism and new classical economics, they are by no means cut from the same cloth. The new classical economists believe that the rational expectations of economic agents will defeat government manipulation of the money supply. Nevertheless, some of their policy prescriptions turn out to be quite similar: reduction of government intervention and uncertainty about government policies, the undesirability of fine-tuning the economy. But their methodology is much different, highly mathematical, and has the characteristic of analyzing the economy as an interdependent whole, in which economic agents understand the consequences of their actions and optimize them in light of their expectations of the future. These economists believe that even in the short run, the economy moves along an equilibrium path, although it is obvious the mathematics of shifting equilibrium is formidable indeed. There is logic to this approach; while not every economic agent is truly rational or acts efficiently on all information available to him, when millions of actors are taken into account, over a period of time, one can begin to see how the economy as a whole might trend as conditions change.

Thus, the neoclassical, neo-Keynesian modeling of the economy is basically retrospective (extrapolation of the past relationships into the future), while the new classical rational expectation models try to provide prospective models, featuring future behavior. It is claimed that their models more nearly fit the real world, where one cannot fool the people all the time. Therefore, such esoteric theory can and has already been translated into practical results. For example, models have been developed using interest rates as a predictor of the economic cycle. Here, the actors are indeed possessed of nearly perfect information—there is no market so free, so full of data, and so quick to respond, now that essentially complete deregulation of financial markets has occurred. In effect, the model embodies all kinds of expectations, even oil price and other "shocks." Before 1979, these models and theory made few initial assumptions: individual preferences of consumers exist, technology is given, and so is the money supply. But in 1979 the Federal Reserve began to control the money supply and allowed interest rates to vary, making these rates a sensitive leading indicator. Thus a new classical rational expectations model employing interest rates as the exogenous variable should, under today's conditions predict future economic cyclical trends very well. Looking backward, such a model developed by H.C. Wainright & Co. (Norris 1985), accounts for the 1982 recession. It is now predicting high growth in 1986, but can it take adequate account of the negative effects of the huge international trade deficit?

Based on such encouraging results, rational expectations lead one to expect economic incentives and disincentives to work very well in the long run, because that is how economic agents think and act. A recent lengthy paper by Laurence Kotlikoff of Boston University (Kotlikoff 1984) shows that new classical models suggest governments can significantly influence the accumulation of national wealth, savings, and investment by adopting appropriate tax policies and taking care not to examine only the short-term effects, but particularly to study the probable long-term results. Although Kotlikoff concludes with cautious caveats about the imperfection of such models, and some other economists could also accept this view, the distinctive feature of new classical economics is that economic agents are consistent and successful optimizers, to the limits of the information available to them. Economic agents do not make persistent mistakes; but their mistakes may nonetheless have persistent consequences, even when the agents are acting optimally! This would account for many of the ups and downs, the lags, actually experienced in the real world. But, in time, markets clear and equilibrium is established. It is true that many psychological and preference factors also influence decisions, but over any reasonably long run, this is the governing characteristic of the economy. This theory states that economic agents are subject to error arising from their ignorance (imperfect knowledge) of the future. This ignorance can also be characterized as risk. Over any length of time or over many independent agents, the errors that result from risk take on a pattern that may be summarized in an objective probability distribution. One can insure against such errors. Thus the rational expectation hypothesis identifies the objective probability distribution of forecast errors with the agent's own subjective assessments.

Uncertainty, on the other hand, affords no opportunity for economic reasoning; in a fully specified model, it is a residual about which one knows nothing. In the long run, it disappears. This is why uncertainty about government policies is so unsettling to risk-taking and innovation; and the frequency with which government policies have been changing in recent years is especially deleterious to long-term growth. Indeed, if government "agents" are optimizers also, as this theory would maintain, their rules for optimization are very different, of short-term horizon, and frequently clash with the behavior of private agents.

But major technological change, the primary residual, also produces such uncertainty; and the models cannot predict it or its effects. While the main focus of the new classical theory (as is true also of other mainstream economics today) has been on business cycle problems, its approach is probably even better adaptable to problems of long-run growth. But it cannot be successful, without studying *how* and *why* firms and individuals make decisions, develop new technology, and change the equilibrium of the whole economy. In brief, technology must become an *endogenous* variable. Although Burton Klein's work (see below), and Schumpeter's before him, do relate innovation and entrepreneurial risk-taking with the process of endogenizing technical change, much more needs to be done; for the force which is the primary cause of growth to remain outside the economic system—*exogenous*—as it has been up to now is unthinkable if economics is to regain any predictive credibility. In this context, the role of capital formation and technical change is obviously central. The Harvard University Program on Technology and Economic Policy directed by Dale Jorgenson and the author will concentrate on these issues. Stanford

University and the University of Pennsylvania are other institutions moving in this direction.

In their thinking about economic agents as optimizers, not passive agents, and in the interrelatedness of the parts with the whole, then, the new classical economists erect their macroeconomics from microeconomic foundations, in contrast to the neoclassical neo-Keynesians. They place great emphasis on theoretical rigor with models. However, the theory is very far from a complete description about how the real economy operates. It assumes equilibrium in a closed system free from outside shocks, and, as noted, it assumes constant technology. It assumes prices clear instantaneously in an "auction" process. The actuality is far more complex and would probably be beyond today's mathematics and empirical findings to handle; nevertheless, these simplifying assumptions have been employed to derive useful predictions and models.

NO ECONOMIC THEORY PROVIDES FOR ENTREPRENEURSHIP AND INNOVATION

The greatest weakness of any of the schools of economics described is that they have no place for innovation, entrepreneurship, or dynamic competition among firms and nations. Yet it is obvious from actual international and historical comparisons that these factors cannot be ignored. Technology and innovation are constantly changing, particularly under the pressure of competition and motivated by individuals and firms which strive to optimize their profits under such pressure. This is the real economy of commodities, work, and risks, as well as the symbolic economy of money and credit. Economists have only just started to think about these processes which businessmen and technologists appreciate from personal experience.

A recent book by Burton Klein of the California Institute of Technology is a good illustration of the search for ultimate understanding necessary for the proper erection of a suitable microeconomic foundation to macroeconomics (see Klein 1984). In the dynamic, constantly shifting equilibrium brought about by firms and individuals competing for growth and profits, the new classical economics' equilibrium is a moving target. Klein, by his own admission an "unrepentant Keynesian," in fact concludes that the greater the propensity for risk-taking, and therefore instability of outlook at the firm level, goaded by competition from domestic and international companies, the greater will be the macroeconomic stability of the country, because firms that succeed in innovation by this process raise their productivity and product quality enough to gain competitive advantage while sustaining moderate or falling prices, and this in turn affects wage demands in those industries, helping to restrain inflation. On the contrary, where innovation and competition are low (as in the manufacture of steel, or automobiles until the Japanese appeared), prices and wages tend to rise excessively, because the demand is inelastic. This in turn puts a deadweight loss on many other sectors of the economy; cyclical recessions occur. By studying data for 387 manufacturing industries, Klein found that the more widespread the rivalry among firms, industries, and across national boundaries, the more firms will constantly innovate and improve (or disappear) and the per capita standard of all the people will rise accordingly. It is reckoning with this pressure of competition (which is not uniform between industries and even types of firms within any industry) that would aid new classical models to more nearly approximate the real world. It is also obvious, therefore, why most economists oppose "protectionism," which they see as primarily hurting a country's own exporters and its internal inflation rate. Unfortunately, the principal competitor of the United States, Japan, does not follow such economists' advice—it imports only about as much as Switzerland. The United States is not without its own devices: measures such as trigger prices, automobile import quotas, multifiber agreements, and agricultural subsidies are scarcely free-market oriented.

The most powerful long-range effects, of course, are found in breakthrough innovations that create new markets. Such innovations cannot be predicted by probability calculations. They are shocks of a favorable kind which, by creating a loophole in the equilibrium of supply and demand, shift the equilibrium of the entire economy in the direction of greater wealth. Such work by an economist is valuable and necessary, even though it does not take into account the cycles in scientific and technological progress which do limit the ability of firms to innovate at certain times, nor does it reflect the changing economic climate in the United States and abroad. (See the author's chapter "The Innovative Milieu" in Lundstedt and Colglazier 1982.)

In light of the foregoing discussion, the large scale experiment with the capital gains tax can be reexamined. This event suggested that perhaps still other microeconomic scenarios might work better toward optimizing the private economy than those hitherto practiced, to improve innovation, productivity, and competitiveness, and from such considerations arose what has been called *supply-side economics,* although this is a misnomer since classical economics always requires that supply and demand both be considered. In its Laffer version, it is a gross oversimplification of a very complex interaction, in which the whole economy participates. But attention to the supply factors in the economy does need to be more extensively applied, and can no longer be ignored.

President Reagan's tax and other policies indeed sought to give more emphasis to incentives and macroeconomic policies that would improve the performance of the U.S. economy, while reducing direct intervention therein. As Martin Feldstein (1981), former Chairman of the Council of Economic Advisers, once put it, "Within the economics profession, there is a widespread retreat from the Keynesian notion that the way to raise income is to expand demand (by government spending). Instead of Keynesian demand management, the new view . . . the essence of supply-side economics . . . focuses on capacity creation through capital formation, research, and innovation." In press reports he deplored the bad name supply-side economics got from extreme rhetoric about self-financing tax cuts and euphoric forecasts of a painless transition to rapid but inflation-free growth. Nevertheless, while not wholly conclusive, the increase in the capital gains taxes actually paid after the 1978 reduction does lend significant support to the concept that some kinds of tax cuts can indeed be self-financing, in that a dynamic competitive economy made up of consistent optimizers will produce growth. Static models of the economy, still widely used today, particularly by the Treasury, incredibly take no account of the future behavior of firms and individuals. Not surprisingly, Professor Feldstein was the only academic economist who publicly supported the capital gains tax reduction of 1978, and he still maintains that it should be a key feature of the tax code.

One of the difficulties in reaching a better understanding of the way the economy could be optimized is that economists find it difficult to talk and work with their theoretical or ideological adversaries, as a recent book, *Conversations with Economists* by Arjo Klamer (1983) vividly portrays. Hence, despite the evidence of how economists build on one another's work (in time), the major dilemma in economic policy today is that there is not yet a discernible consensus among economists on a general set of macroeconomic policies (fiscal and monetary) which would provide for cyclical recovery and satisfactory long-term performance. It almost seems as if most economists have abandoned striving for understanding of the larger picture. At the December 1984 convention of the American Economic Association, only one of over four hundred papers focused on the federal budget deficit! This is most regrettable since, as economist Alan Greenspan once said about his Washington days, all important changes in government policy originated in academia.

Absent such a consensus, governments continue to seek to intervene in the microeconomy as political pressures of the moment dictate, hoping, as Keynes did, to produce short-term benefits. There is still an inadequate understanding of the set of opportunities and calculations involved in developing and introducing new technologies, which fuel growth. (As mentioned, the new classical economists take technology as a given; future work must address how their models can incorporate such provisions for changing technologies, perhaps employing some of the useful results of Professor Klein and of the input/output analyses of Nobel Prize winner Wassily Leontief as exemplified in the May 1985 *Scientific American*). Hence, it is difficult to demonstrate how specific macroeconomic and microeconomic policies will necessarily translate into innovative decisions by firms that will result in consistently rising levels of productivity and economic performance generally. Table 1–3 gives examples of the different types of policies.

In 1981, in a book entitled *Taxation, Technology and the U.S. Economy* co-edited by Bruce Hannay and the author, Dale Jorgenson (1981), a prominent Harvard microeconomist, following another approach than Burton Klein, involving disaggregation of the economy into thirty-five industrial sectors, addressed the puzzle of the declining rate of productivity increase in the United States in the seventies, which was nearly zero during the decade. Among the major reasons identified were the decline in technological progress, and insufficient capital formation (both physical and human). This is not inconsistent with the more detailed industry analyses of Klein. Jorgenson's essay urged prompt measures to reduce the taxes on capital to encourage new investment (for as he showed in his correlations it is the after-tax rate of return which governs decisions to innovate or invest and accounts for the changes over time in the different sectors), as well as increased research and development. Recent work by him has extended and confirmed his findings for the critical period from 1973 to 1979 (Jorgenson 1986). The experience of many in the industrial world corroborates this conclusion, as this period coincided with the era of escalating energy prices, increased international competition, declining innovation in some (but not all) industries, and the consequent rapid obsolescence of much of the capital plant in the United States, even though much of it is still in use. The recent report of the Presidential Commission on Industrial

Table 1–3. Some Factors Affecting Technological Innovation and Growth

Macroeconomic Policies (Climate Creation by Governments)	Microeconomic Policies (Direct[a] and Indirect Intervention to Affect Incentive and Performance of Industrial Firms)
Expenditure levels by all layers of government (expressing total burden of government on the private sector) Budget deficits and their timing	Government spending to affect specific components of the private sector, such as R & D, infrastructure construction and maintenance, educational programs, control of capital spending decisions.
Tax policy (including such issues as overall level of GNP preempted government taxation)	Form of taxation, i.e., its specific incidence upon capital formation, entrepreneurship, consumption, etc.
Control of monetary supply	Price and wage controls, price supports and subsidies; "bailouts" of failing companies
Monetization of debt ("printing money!") vs. borrowing from the private sector to finance deficits	Rationing of credit; ceilings on interest rates for certain types of savings; usury laws; deductibility of certain interest costs
Intervention in international currency markets to manipulate value of currency (also relationship to gold)	Favorable financing of exports by Export-Import Bank, tariffs, quotas, etc., on imports Regulatory and antitrust policies Minimum wage laws and other labor union legislation; labor mediation
Free trade policies	Protectionism

a. These fall largely under the heading "industrial policy."

Competitiveness examines all these relationships in great detail; it is clear that improved American competitiveness in those sectors which can benefit from technological progress and good macroeconomic policies *is* the prerequisite for a greater growth rate of the economy. That is the major reason for striving for greater competitiveness by the managers, workers, and entrepreneurs of these sectors. Other sectors, which face less competition from abroad, need to find it domestically, to contribute to such growth, which is the fundamental requirement for a more just and progressive society.

The new classical economics of rational expectations has forced governments to understand the need to make their policies credible. What is still lacking (and perhaps is only remotely foreseeable) is the "hardness" of physics in the science of economics, so that the uncertainties at the micro level (as in quantum physics), governed largely by probabilities, and unexpected actions, can convert into satisfactorily predictable likelihoods at the macro level (akin to classical Newtonian physics). But it is clear that the new classical economics sees this need also and is attempting to address it in an increasingly sophisticated way. It is nevertheless certain that future evolution will incorporate new findings and new concepts, and other schools of economics. The Nobel Prize economist, Herbert Simon (1985), underlined the situation as follows: "Existing uncertainties about the correct explanations for economic growth and business cycles cannot be settled by aggregative analysis within the neoclassical framework. Current disputes in theory rest largely on ad hoc, casually empirical assumptions about departures from perfect rationality under uncertainty. Such disputes can only be settled by painstaking microeconomic empirical study of human decision making and problem solving." In short, Simon recommends disaggregation to the firm and individual level, where technology, psychology, and other motivations come together and are made *endogenous* within the economic system as it moves from one equilibrium path to another.

Because of these theoretical limitations, one must proceed very cautiously with new policy initiatives, particularly those not based on hard empirical evidence from actual businesses and enterprises. Such a case is the capital gains tax macroexperiment, which showed that with little *downside* risk, because the possible revenue effects were small, large positive benefits could be realized. The reverse situation exemplified by the 1985 proposals (discussed later in this chapter) should be sedulously avoided. However, while economics may have many limitations, it has also gained many quantitative insights and methods, which must be given time to catch up with reality. It makes no more sense to say that the perceived shortcomings of macroeconomics mean that all the postwar work is false and should be jettisoned as to declare that quantum physics rendered Newtonian physics incorrect. Rather, it revealed its incompleteness and led to further fruitful research which extended the scope of all physics.

"HIGH TECH" AND TECHNOLOGY

In the short time since 1981, the political and economic climate has changed very materially. Strong recovery with low inflation is here, but so are high real interest rates and $200 billion domestic deficits, with large trade deficits. Once more, the country is in danger of forgetting the fundamental lessons of the postwar era. It is still living beyond its means. There is thus a new juncture in the understanding of technology and the state of the economy. High-technology industry is regarded by many as the way to maintain economic growth. "High tech" is the buzzword of the moment. There is a need to ensure that it is not a passing fashion but a permanent focus of attention and action. Yet, there is no quick fix for the economy in high technology. A favorable environment for engineering, technology, and innovation is a long-term necessity. It takes many years in some cases to move a new invention into full commercial operation, and longer to make it really economic, moving it along the learning curve. Technologists and economists have an obligation to make sure that the political community understands that the innovative private sector will require substantial but steady attention over many years. Fine-tuning is simply impossible in the complex American economy. *Managements and entrepreneurs will have enough rapidity of change from technology and competition that they will increasingly be unable to cope with constant rapidly changing intervention in the economic sphere by political authorities.* Recently, an experienced French executive, A. Chalandon, who knew at first hand what governments can do, said that "government has a shorter term view of things, whereas decisions in industries like oil are taken with an eye to the next 15–20 years." From "rational expectations" theory as well as experience, could one expect any other situation? Richard Cooper (*New York Times* 1984) of Harvard University recently expressed the same viewpoint, urging retention of present policies for a much longer period of time.

What is the sense of the political climate toward technology now? President Reagan's speeches during the 1984 election campaign were strongly supportive of technology and incentives for it. In September 1984, the president said,

> Let's remember that technology is born from capital, and capital requires incentives for risk-taking and investment. We've seen dramatic proof of this relationship. The 1978 capital gains reduction, followed by our 25% reduction in the tax rates and the shortening of depreciation schedules for business investment, have given an enormous shot in the arm for risk-taking and entrepreneurship in America. (*New York Times* 1984.)

In his State of the Union Address of February 1985 and the award of the first round of National Medals of Technology he continued this emphasis. His April 1985 speech at St. John's University and subsequent speeches concerning tax reform called this "the age of the entrepreneurs." In the Soviet Union, the new leader, Mikhail Gorbachev, has recently pleaded for more innovation, research and development, efficiency, and progress at a meeting on science and technology of the Central Committee of the Communist Party.

To many people high technology means just computers and electronics. Yet there are many more industries that are technology-based: chemicals, pharmaceuticals, textiles, nonferrous metals, agriculture, machine tools, aircraft and aerospace, automobile manufacture, energy production. They experience increasing international competition and must innovate more strenuously than ever before to survive and prosper. Jorgenson (1981) concluded that "declines in rates of technical change for *individual industrial sectors* of the U.S. economy must bear the full burden of explaining the slow-down in the rate of technical change for the economy as a whole." As a consequence, he advocated that "the major focus for economic policy should be to stimulate the development of new technology for *all* industries" (emphasis added).

What is in fact occurring is also a major thrust of technology into the service sector of the U.S. economy, which will significantly improve its productivity, while the more traditional industries are modernizing and automating, or shutting down their plants. The United States is experiencing the greatest technological revolution in at least fifty years as the information and communications age spreads throughout the industrial world, replacing in part the old energy- and material-intensive industries; already the capital invested per information worker (half the working population) is equal to that invested per manufacturing worker (about one-fifth of the workforce). There is thus a very large unrealized potential from innovations already known (similar to the situation during the 1930s); the new innovations will be mostly for the twenty-first century. The actual capital requirements to exploit such potential and realizable technology will vary from industry to industry, but if the growth rate is to be raised, the totals will surely be larger than ever and *raising the growth rate of the United States is the real agenda of the 1980s and 1990s.* With a declining birthrate, this goal requires higher productivity increases. It can be foreseen in the service industries by the continuing spread of computers and telecommunications, and in manufacturing by more rapid adoption of robotics and changed working conditions. In both cases the level of education of the workforce needs to be raised, and technological literacy more widespread.

These requirements for greater growth obviously will necessitate new public investment as well, particularly in human capital. Thus the "dismal science" of economics (which is the science of the allocation of scarce resources) will return, despite great technological advances, because there is a fundamental balance required among conflicting demands: human and capital investment for greater growth versus present greater consumption.

The science and technology now available and in prospect make possible the spawning of many high-, medium-, and low-technology businesses that provide jobs, creativity, and competitiveness because the entry price for new firms desiring to utilize these technologies is much lower than for the smokestack industries. With more entries there is more competition, more propensity to engage in risk, more failures, more innovative activity and greater growth and stability at the macroeconomic level. This climate of opportunity explains why entrepreneurship and the entrepreneur are the keys to growth in the rest of this century and beyond. Even Socialist Sweden is changing its attitude toward entrepreneurs. As one of its economists said recently, "For 40 or 50 years, building your own business had been seen as little better than being a gangster." Only the Communist nations still make enterpreneurs illegal!

The years ahead are sure to be both wrenching and fascinating.

TECHNOLOGY AND EMPLOYMENT

Because technological development and the economic framework in which it takes place have strong feedbacks, the essence of a dynamic economy, it is important for engineers and scientists to work together with economists to inform the governmental decision process. The issues are complex. Take, for example, the impact of technology on employment. Civilian unemployment, now around 7.0 percent, will again dominate the political discussions as soon as it starts to increase.

There are many reasons why unemployment (especially its "structural" or "natural" component) is likely to remain higher than was once considered minimally acceptable, and a greater microeconomic understanding will be required before it can be dealt with adequately. There has been a major increase of women in the workforce. The postwar baby boom has just crested. Many industries are losing their international competitiveness and may well be beyond full restoration as the economy shifts to a different mix of goods and services. The secondary educational system has deteriorated. Minimum wage and welfare policies have tended to make work in the lower paying jobs unrewarding. The high tax structure coupled with the inflation that prevailed until only recently has forced many people into the underground or cash economy, where they are considered unemployed. The heavy illegal immigration has displaced Americans from hard low-paying work that was and is available. There have been important sociological changes in American families. Matching of desired jobs with job seekers, particularly at excessively high wage levels for some workers, is a very challenging task. All of these factors and others contribute to what monetarists and new classical economists, at least, consider "natural unemployment," but that does not say the level cannot be affected by the right microeconomic policies. Competition and innovation may be better, more humane methods of reducing unemployment than harsh monetary shocks.

Yet, it must also be realized that the firm instability created by competition also creates instability for the individual employee—the price paid for greater stability and growth at the macroeconomic level. High-technology industry has been advocated as a possible way out of this national problem. But high-technology industry may eliminate large numbers of traditional jobs, and in any event accounts for only 10 to 15 percent of new jobs, the rest coming from traditional low- and medium-tech industries. Many companies, unions, and government units are combatting this trend by extensive retraining programs. Educational needs for training the youth and updating practitioners for an increasingly technological society are being strongly addressed in many quarters. Education is an investment in human capital and must be greatly strengthened for the information age. The unique American system of research universities and their productive linkage to government and industry requires much greater support from the public.

It is important to be sensitive to the fears of those now employed but who are concerned about future employment in a higher technology society. The problem is very real. Can the worker who is puddling steel become an assembler of silicon chips on an assembly line? It is not the same kind of work, and the pay is different. Even though the population growth has slowed, jobs are certain to be a key public policy issue for the rest of this century.

Recently some important statistics bearing on the subject have appeared. In 1979 David Birch (with Susan MacCracken) of MIT completed a study based on 5.5 million companies in the Dun and

Bradstreet files (Richman 1983; Levinson 1985). Looking at the decade of the seventies, they found that the U.S. economy had generated about 19 million new jobs, a quite respectable performance in an era when the baby boom cohorts were entering the workforce in large numbers. It was a much better performance than that of Western Europe. But where did the jobs come from? Seventy percent or more came from small companies of fewer than one hundred employees; most of the rest from government, primarily at the state and local levels; very few from the Fortune 500 companies. This astonishing discovery received very slow recognition, although many large company executives in the United States and Europe recognize its basic validity.

Birch's latest research reasserts that it is the young, small companies that are the sources for most jobs. He points out that, in particular, it is the creation of new companies which is the key requirement for job growth. Of course, many companies die; and new ones are born, but out of the new small companies there will emerge a few which will experience explosive growth, graduate into the large-company category, and create many of the new jobs in the future (as did IBM, Xerox, and the like). Thus, it is the climate for formation and growth of entrepreneurial new ventures (whether primarily technological or not) which is crucial; and it is different from the environment required to nurture ongoing small businesses, and that is different yet from the needs of large business, and none of these can be neglected for a proper growth strategy or jobs strategy. It is also important to note that small business is the major training ground for new workers, many of whom ultimately go to large corporations; there is also a reverse flow as entrepreneurs leave to found their own businesses.

Birch's findings for job formation in the seventies are not inconsistent with the data embodied in Figures 1–2 and 1–3. The jobs so created came to a large degree from companies that had been founded in the sixties, and very probably the companies that will produce most of the new jobs twenty years from now do not yet exist. One clue is offered by a Joint Economic Committee report that traces the venture capital process in the seventies and tracks the results to the end of the decade. Out of 1,332 companies, 72 public companies were selected. By 1979, on a venture capital investment of $200 million, these 72 had created 130,000 jobs. At present, over ten times more venture capital is being invested annually. If these results (although by no means random) are reflected in the much greater number of new companies being formed in the eighties, even allowing for a low net success rate would still result in millions of jobs. While some existing jobs may also be lost as a consequence, history shows that, over time, the net gains are real and long-lasting. Thus, agricultural employment has diminished in one hundred years from half the workforce to slightly over 3 percent, but the far larger number of jobseekers were accommodated in a growing manufacturing sector, which reached 33 1/3 percent of those employed before beginning to drop toward the slightly over 20 percent of today. By the end of the century it will perhaps be only 10 percent, but the jobs are increasingly found in the service and information industries. Nevertheless, because most services are not exported, the United States must always maintain a healthy manufacturing sector to remain competitive in international trade, which is a growing part of the economy. Furthermore, without the purchases by manufacturers and certain service companies, the high-technology companies would have no domestic customers.

Venture capital by itself is insufficient to fund the roughly 600,000 companies started annually in the United States (today's rate of start-ups is seven times greater than the start-up rate for the 1950s and 1960s). Much of the venture capital comes from relatives, friends, and personal savings. But venture capital is crucial to the formation of the ultimately explosive type of company. All the investors eventually desire a market from their holdings so that overall market liquidity is essential. This is especially important for the entrepreneurs and employees of new companies, for whom founder's stock is the key incentive to pull them away from secure jobs. Subsequent financing by public issues, and "bridge" funding and "mezzanine" funding, are required for further growth in most cases. Thus, the entire equities markets' health is a prerequisite for the growth of companies and jobs.

MANAGEMENT AND LABOR

What about American management and labor in all this? It seems evident that a country so dependent on technology as the United States must have managements in which technologists play an important role at the highest levels. It has not always been so, and particularly in industries most affected by inflation, where the financial expert (often an MBA) has tended to dominate. That era is now receding; inflation no longer covers up management failings, and "hands-on" managers and changed organizational styles are increasingly needed by the larger companies. (In Forbes' twenty top corporations, in all fields, nine of their chief executive officers are technically trained.) In many technologically strong companies, this financial era never existed; and managements continue to focus on long-run competitiveness and growth, even though their equities values often gyrate wildly as a consequence of the very narrow horizons of institutional investment managers (hypnotized by apparently high real interest rates and uncertain monetary policies, which keep stocks from rising with a growing economy). Many large company stocks traded on the New York Stock Exchange have a heavy participation by such institutions. The American Stock Exchange is less so dominated, and over-the-counter trading even less; and here is where the entrepreneurial growth companies usually go public first, and where stocks are traded freely, thus assuring the initial risk-taking investor of ultimate liquidity. Recently the British have introduced a form of limited public financing (the Unlisted Securities Market) which is much less complex than the U.S. securities laws require. A similar financial innovation is perhaps needed in this country.

Organized labor, which now represents less than 20 percent of the workforce, must also grasp the implications of changing technology. Have past policies contributed to a rise in income for some workers at the expense of unemployment for others and loss of international competitiveness which leads to still more loss of jobs? There is embedded in this question a clear new tradeoff, *employment versus wage levels,* to add to the formerly well-known inflation/employment trade-off. *The Economist* (1985) plotted a curve of the rise in unemployment rates from 1975 to 1982 as compared with wage rigidity (a measure of the degree to which wages respond more to prices than to unemployment). This curve shows Britain, France, and Holland at the high end, West Germany and the United States near the low end, and the Japanese lowest of all. Management and labor are recognizing that a high-wage economy requires wages that reflect company profits, based on competition, and a very high capital investment per worker; hence national policies must encourage human and physical capital formation. New capital frequently embodies new technology which improves productivity. Microeconomics must focus on all these issues.

Based on recent experience and the effects of international competition, then, American managements appear to be responding more rapidly than labor, for example, by increasing R&D and capital spending in real terms and improving their breakeven points by ruthless cost-cutting. This welcome adaptation to changing circumstances is still largely confined to the manufacturing sector, where R&D and growing productivity increase are encountered (all technologically based industries together account for 75 percent of total R&D private funding, which is increasing at 8 percent per year). The nonmanufacturing sector except for agriculture performs relatively little R&D and acquires its technology (such as computers) from outside sources. As a consequence of this and the relative lack of overseas competition, productivity increases in the nonfarm, nonmanufacturing sector have been much smaller since 1970. Financial services and communications are significant exceptions. Agriculture is a high-tech industry with a large entrepreneurial history and content (both public and private), and has been doing very well in productivity gains.

THE CORRECT CLIMATE FOR TECHNOLOGICAL CHANGE AND ECONOMIC GROWTH

Growth propelled by technological change requires a sound climate which presupposes correct government policies. The federal budget deficit and its effect on growth policy is a matter of great concern in this connection, tending to obscure the economic fact that the total burden of government on the private sector is measured by its spending, not by how that is financed (although the actual methods adopted can have important political and economic effects). Nevertheless, deficits can mean higher interest rates and probably a cutback in credit demand by the interest-sensitive sectors of the economy, such as housing and automobile manufacture, as well as diminished exports. The deficits themselves have been inherently stimulating, and helped initially to reliquify the balance sheets of many American businesses and consumers. The great danger is that constructive capital formation, which embodies much new technology, will be impeded by the high apparent real after-tax interest rates (Figure 1—5) now prevailing, reflecting investor expectations about the future. These rates are apparent in that they represent correction of nominal interest rates for the *prevailing* inflation rate; but investors' expectations of inflation and risk may be higher than these, meaning that the real expected rate of interest may be lower than these curves show. David Ranson (1985) points out that British inflation-indexed bonds suggest a real rate of interest of 3 percent, implying high investor inflation and risk expectations. But these high rates do serve to improve savings incentives in the long run, channel capital into the growth, more technologically based, competitive or "sunrise" industries of the future, and force businesses to be much

Figure 1–5. "Real" Interest Rates.

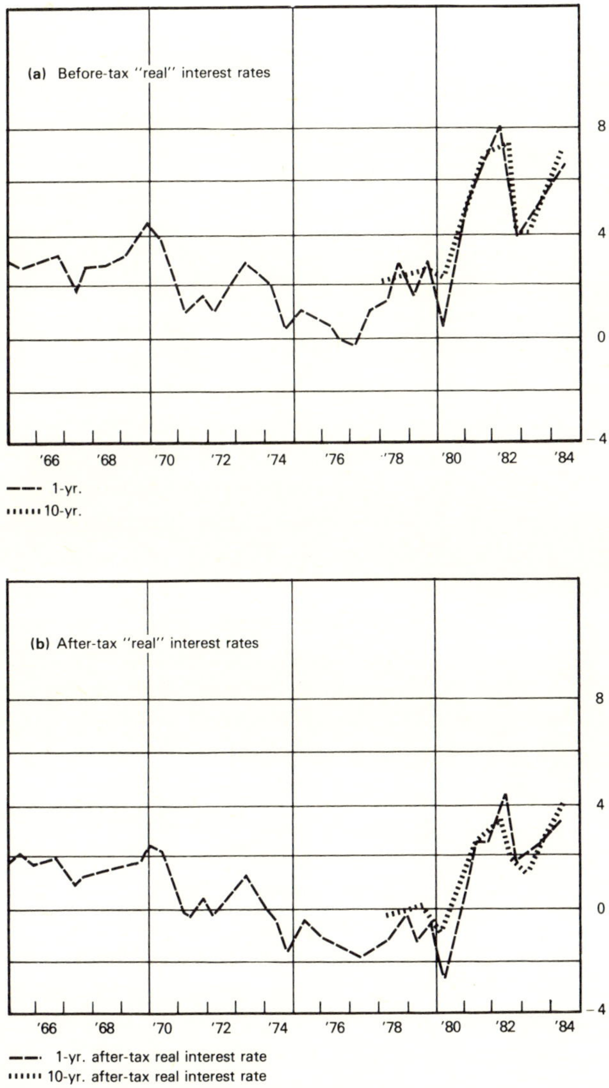

Source: Federal Reserve Bank of St. Louis.

more prudent in their inventory management, a factor which may well reduce cyclical economic variations. And the concern for deficits has imposed a sort of financial discipline upon politicians, based on the rational expectations of investors—a new factor to the international political economy.

It is probable that real interest rates will remain higher than in the past, but current microeconomic understanding of all the factors that influence them is too imperfect to venture any quantitative predictions. A.S. Holland (1984) of the St. Louis Federal Reserve Bank has recently studied the real interest rate phenomenon, and concludes that variability in the money growth increased economic uncertainty and the risk premium on interest rates. The expectations of future government deficits seem also to have contributed. Nevertheless, excessively high apparent real interest rates mean high "hurdle" rates for investment by companies and individuals, and a high cost of capital. This can only lead at the end of the day to the deindustrialization of the United States and a continuing export of jobs, even with excellent innovations and rivalry. It is also threatening the solvency of much of the U.S. farm sector, staggering under a massive debt burden as well as the effects of the hard dollar (although much of the fault lies also in unwise government intervention in the form of price supports and subsidies, as well as production controls, instead of allowing the traditional market forces greater reign).

A major consequence of the recent macroeconomic policy, however, has been this strength of the dollar. Whereas in 1981 the United States still had a favorable current account balance of payments, in 1984 it will have been unfavorable by perhaps over $100 billion. Many American companies and industries have been adversely affected, even extinguished; and perhaps two million jobs have been lost, although others have been created. Even innovative companies are affected. It is true that there are positive elements to the hard dollar: competition from imports helps restrain inflation and forces greater innovation; capital flows into the United States from abroad to augment domestic savings; and the economies of other nations are being revived by their growing exports to the United States (there is no indication that they are suffering from capital shortages as a result of the capital flows to the United States). In theory, such an unfavorable trade balance should eventually work toward a weakening of the dollar in a self-adjusting mechanism; but it is not occurring rapidly as yet. At its peak, the dollar was overvalued by perhaps 35 percent; if long continued, some industries will be permanently lost to the United States, including those of great importance. There is already an unfavorable balance of trade in high-tech electronics, and much of the recent capital spending boom in the United States has come from imported equipment. American manufacturers are being compelled to move some of their production plants abroad, while foreign manufacturers build plants in the United States to protect their growing American markets against future protectionist actions.

The DuPont Economics Department has prepared Table 1–4, which shows the change in American manufacturers' competitiveness as a result of productivity changes, labor rates, and currency adjustments. The left-hand section shows the effect of man-hour productivity and wage rates on labor costs in several major countries for 1980–1984. The right-hand section shows the currency effect for the same countries in the same period. The far-right column combines the three effects.

While the United States has made significant gains in productivity, it has fallen sharply behind Japan—a challenge to management. Likewise, management and labor have a role to play in wage-rate restraint. Using local currencies, and combining these two effects, the column "unit labor cost" shows that the United States still lags behind Germany and Japan. However, taking this effect and translating through currency values, the picture changes drastically and the hard dollar (beyond the influence of management and labor) becomes the overriding factor in relative competitiveness. The United States is noncompetitive to all the other countries, and the principal European powers come out the leaders! They thus are seen to have little incentive to help reduce the grossly overvalued dollar, unless faced with rising protectionist pressures in the United States. It is perfectly clear that the real challenge is to American economic policy, but then the other industrial countries have a responsibility to stimulate their economies, if a trade war is to be averted. Recent currency intervention by the Group of 5 is having an effect.

There are surely other factors at work, including the preference of foreign investors for the higher interest rates and the safety and security of U.S. free enterprise policies, insulated from political unrest. The decline in foreign investment and lending by American companies and financial insitutions, despite the favorable dollar for such purposes, is also a significant cause, and demonstrates the attractiveness of the United States, as well as the incentive effects of the 1981 investment tax credits. These factors are inherent in international capital flows, which are now $20–30 trillion per year, dwarfing international trade in goods and services at about $2 trillion per year.

Table 1–4. Comparison of Dollar Unit Labor Cost—U.S. versus Key Industrial Countries.

	Unit Labor Cost Effect				Currency Effect				
	1980	1984	% Change 1980–84	Unit Labor Cost (1984 as % of 1980)	Dollar Value of Foreign Currency[a] 1980	Dollar Value of Foreign Currency[a] 1984	% Change 1980–84	Unit Cost Based on Currency (1984 as % of 1980)	Combined Unit Labor Cost[b] in Dollars (1984 as % of 1980)
France				141.5	.237	.114	−51.9	48.1	68.1
Productivity	1.123	1.327	18.1						
Wage rates	148.1	247.7	67.3						
Unit labor cost	131.9	186.7	41.5						
Germany				107.0	.55	.351	−36.2	63.8	68.3
Productivity	1.088	1.221	12.2						
Wage rates	125.0	150.0	20.0						
Unit labor cost	114.9	122.9	7.0						
Italy				177.5	.00117	.0006	−48.7	51.3	91.1
Productivity	1.169	1.285	9.9						
Wage rates	160.2	312.5	95.1						
Unit labor cost	137.0	243.2	77.5						
United Kingdom				119.4	2.326	1.336	−42.6	57.4	68.5
Productivity	1.013	1.192	17.7						
Wage rates	162.8	228.7	40.5						
Unit labor cost	160.7	191.9	19.4						
Japan				95.2	.0044	.0042	−5.0	95.0	90.4
Productivity	1.286	1.634	27.1						
Wage rates	121.2	146.6	21.0						
Unit labor cost	94.2	89.7	−4.8						
United States				113.1	1.00	1.00	—	100.0	113.1
Productivity	1.017	115.6	13.7						
Wage rates	132.7	169.3	27.6						
Unit labor cost	129.5	146.5	13.1						

Notes:

a. Yearly average exchange rates.

b. Based on exchange rate movements through Fall 1985, the unit-labor-cost differentials between the U.S. and other countries shown in the last column would still be significant although somewhat smaller.

Wage rates indices 1977=100.

The consequence is that the trade deficit has weakened the stimulative effects of the domestic deficit, and growth has slowed. In addition, the large deficits also add new interest charges to the federal budget every year, a process that cannot continue indefinitely without grave economic consequences. As a result, the current political debate is addressing the presumed links between the deficits, high apparent real interest rates, the hard dollar, job losses, fragile international credit arrangements, and threat of a new recession.

Yet one must maintain a long-term historical view of what is happening in the industrialized countries. Unlike the sixties and seventies, but perhaps more like the twenties, the major trend is disinflationary and more internationally competitive; with it, a declining longer term trend in interest rates. Although people do have rational expectations, there is a time lag during which perceptions of the new reality slowly take hold; this is clearly a volatile world in the short term. It is also apparent that governments and economic agents have learned much from the past: deregulation has been proceeding; the Organization of Petroleum Exporting Countries (OPEC) is weak; the monetary authorities are more cautious. If the members of the Organization for Economic Cooperation and Development (OECD) can manage the transition from this period to one of longer term low inflation without a financial crash or domestic recession, there is a real possibility of lower interest rates and genuine sustained economic growth—an exciting prospect, indeed. That would certainly be the right climate for technological innovation and entrepreneurship.

But how is this to be done? One very important guide to augment the concrete historical experience in the United States is to look abroad.

The Japanese successes are much more due to their economic and labor policies and the high savings rate than to any central planning by Japan's Ministry of International Trade and Industry (MITI), as a closer study of their history will show. The key to Japan's growth has been a very close cooperation among government, business, and the financial community, not as specific intervention to pick winners and losers. The latter function is performed by the business community operating under a very favorable macroeconomic policy of low inflation, low interest (and low cost of capital, probably about one-half that of the United States), and favorable regulatory and tax policies. Nevertheless, their former budget deficits were beginning to pose problems for them not unlike those in the United States, if at a different level. It is noteworthy, however, that Japanese entrepreneurs like Honda and Sony have triumphed worldwide despite MITI; while many industries such as aluminum, chemicals, agriculture, pulp and paper, and shipbuilding are weakening. Japanese wage policies, for example, entailing large bonuses if their companies are profitable but lower fixed wages, are less rigid than those of other nations, and have much to recommend them. As a recent *New York Times* (1985a) headline said: "Labor Peace Reigns in Japan, and Unions Wither." Indeed, the Japanese experience points the way to greater participation and satisfaction of workers in their jobs. Martin Weitzman of the Massachusetts Institute of Technology Economics Department has recently published an important book (1984) on this subject which may well receive more attention as the next recession arrives. Richard Nelson (1985) of Yale in another recent book says: "If MITI does not seem likely in our future, the flexible industrial structure of the United States should not be discounted as a formidable competitive engine of progress. We may be lucky that it so stubbornly resists being targeted, coordinated or planned."

Another guide is to be found in Europe. The data cited in Table 1–2 on employment are the key to why the principal subject on the minds of European policymakers is unemployment, with great concern for the effect of technology on jobs and international competitiveness.

In looking behind the startling differences between Europe and the United States, one may find the following possible reasons and discern more clearly what should be the path for U.S. policy in the future:

1. The macroeconomic and microeconomic policies of the European Economic Community (EEC) countries have set different equilibrium conditions between capital and labor during these years. Although U.S. productivity fell relative to Europe's in the 1970s because of these different policies, the United States still has the highest productivity in absolute terms. The productivity differences are very great between different industries, as noted by Burton Klein and by Dale Jorgenson, so that one must be careful in defining the reasons for the differing rates of productivity increase. Furthermore, productivity statistics cannot quantify the fact that product qualities also change with time (as the Japanese experience proves). Fundamentally, it has been cheaper to use labor than to add capital in the United States, at least for the last twenty years, thus favoring labor-intensive businesses. Jorgenson has shown that the growth of the U.S. labor supply in the period of 1948–1973 was 1.6 percent per year, but jumped to 2.7 percent in 1973–1979 (because of entry of members of the baby boom cohort into the labor market). But

capital input averaged increases of 4.3 percent versus 3.8 percent in these two periods; the workforce increased, while the investment rate was decreasing—the capital/labor ratio grew a negative 1.6 percent during 1973–1979. Inflation had a great deal to do with this, as did the oil shock. Wage rates have fallen or remained static in real terms during much of this period, whereas there has been an increase in real wages in the EEC. Consequently, the Europeans invested in capital to reduce costs at a higher proportion of GNP than in the United States rather than in new capacity, and their productivity increases rose above those of the United States. European social policies, and stronger unionization and labor rigidity, have also contributed to this mix. Nevertheless, the net result has been declining growth rates of their economies, as international competition and technological change accelerated with rising unemployment. These differing effects illustrate how difficult it is for governments to get the policies for growth with low unemployment just right.

2. Both macro- and microeconomic policies also have a severe effect on currency relationships, and these have a profound impact on the export balances. The large market in the United States is a much more secure base for American companies than the still fragmented national markets of the EEC, limited by national nontariff barriers and other attitudes.
3. America's policies in the economics sphere and its risk-taking culture have tended, relative to Europe, to favor smaller companies and their formation. It is true that, while on the whole the largest companies have not added employment, these numbers are net; jobs are being created and destroyed over the time period involved. Small companies are gross generators of jobs, but these must be corrected for the inevitable high rate of failure of young companies. Still, new ones arise every year. As Birch states, it is also the explosive growth of some of the smaller companies which ultimately proves to be a major factor in new job formation. Further examples are McDonald's, Federal Express, or Disney Productions, which illustrate the point that many of the new firms and new jobs do not involve high-technology, but nevertheless contribute to the flourishing of an entrepreneurial environment generally.
4. In the eighties and nineties the United States can increase productivity without creating more unemployment, because the number of workers entering the workforce will decline. Between 1965 and 1979, the American labor force grew at about 2.2 percent annually. Between 1985 and 1995, this rate should average nearer 1.5 percent.

Over 7 million jobs were created in the United States in 1983–1984, and 2–3 million more in 1985 are possible, against new workers of 1.5 million, provided growth of the economy can be maintained at above 3 percent (the postwar average). Unfortunately, this does not seem so likely now, under prevailing policies, and unemployment is barely diminishing; rather, it may start to rise again as the foreign trade deficits increasingly sap the growth of the domestic economy, or when the next cyclical downturn occurs. Shortages of many types of skills will, nevertheless, be increasingly noticeable, making it even clearer that the balance between labor and capital must shift. Hence, this will be a period of receptivity to the application of new technology throughout the economy; and there is need for macro- and microeconomic policies which will favor the formation of new companies and the adoption of new technologies by existing companies, rather than directly protecting declining industries. It is, therefore, apparent that the time for lowering interest rates and taxes on capital is approaching (without disturbing favorable incentives already in place). That is, the burden of government on the private sector, which has been performing these feats that are the envy of the Europeans, should lessen. It is noteworthy that the overall tax burden for OECD Europe was 46 percent in 1982, versus about 32 percent in the United States.

Another way of analyzing the appropriate economic policies is to examine more specifically recent experiences in the United Kingdom and the United States. An extensive treatment of this has been written by Walter Eltis, Fellow, Exeter College, Oxford, for the *Bank Credit Analyst* of July 1984 (Eltis 1984). In sum:

1. The United Kingdom under Prime Minister Thatcher gave priority to reducing its budget deficit: it was cut from 6 percent to 3 percent of national income, and may be 2 percent currently; the U.S. deficit rose from 3 percent to 7 percent, and is now about 5 percent. But to accomplish this while maintaining tight monetary control (to avoid excessive capital outflows), the United Kingdom had to raise taxes (by about 18 percent). Its overall tax and spending burden is much higher than in the United States. The United States cut the tax burden by 4 percent under President Reagan, so that it practiced a tight monetary policy but a loose fiscal (budget) policy. Both countries increased government expenditure.
2. As a result, the United States raised its employment much faster after inflation was broken by monetary correction. But the tighter

U.S. monetary policy has kept interest rates higher, and the equities markets relatively lower (the Dow Jones Industrial Average has risen 30–40 percent compared with 100 percent in the British Share Index). And, of course, the dollar became harder.

3. As the United States is rapidly approaching full employment (above the natural rate, which, has many structural origins), the United States could remain at such a level for a long time if its financial affairs are corrected. It cannot go on borrowing with impunity from successive generations, because they must pay an increasing share of the GNP once the rate of interest becomes higher than the rate of growth. The relationships, and the time intervals, are discussed in detail in Eltis (1984).
4. The United States can create a nearly ideal macroeconomic climate by either cutting expenditures so that the deficit approaches 2 percent of the GNP, or, if unavoidable, by a suitable combination of budget cuts and tax increases. At present, federal spending is 24 percent of GNP. By 1990, it should be reduced to 21 percent. Taxes now 19 percent would not have to be raised, as the economy can tolerate a 2 percent deficit indefinitely. Indeed, if the growth rate of the economy can be kept higher than the growth rate of spending when the 21 percent level has been reached, it should be possible to *reduce* taxes and spending, a most desirable long-range climate indeed! If taxes are to be raised at all, they must not be on incentives to save and invest, and not on income but on consumption. It is clear from what has been said before that tax incentives, such as the capital gains tax reductions, have contributed significantly to American job formation, and innovative competitiveness. It is encouraging that President Reagan in his January 1985 press conference, February 1985 State of the Union Address, and subsequent press conferences continued to oppose tax increases.
5. If such a "soft landing" could be achieved by the United States, world interest rates will indeed fall, and U.S. interest rates even more. The dollar will depreciate, exports will increase, and American competitiveness will improve. Such a gain in sales abroad would at least match any loss in domestic demand brought about by any kind of a moderate increase in consumption tax, or reduction in government spending. The equities markets would rise. The United States would have stopped inflation without the prolonged unemployment experienced by the United Kingdom's more balanced-budget, low-growth approach. This would seem to be a prescription for the ideal macro- and microeconomic policies to favor steady and

higher growth with increasing productivity at low inflation and low unemployment—suitable for innovation and an increasing standard of living. Innovation, as mentioned before, can be strongly affected by incentives and climate.

6. If, on the contrary, American policy becomes stalemated, the continuing high deficit will cause the difficulties mentioned above, and place great temptation before the Federal Reserve to lower interest rates and reduce the value of the dollar, at least in the short run, by "printing money"—monetizing the debt. However, the rational expectations of investors would soon compensate, inflation would rise again, and foreign investors would lose confidence. The dollar would fall precipitously, thus further aiding the inflationary trend. The country would approach the status of a banana republic! The post-election euphoria in Washington was unjustified; the United States is pushing its luck if it persists on its recent course.
7. The United Kingdom, on the other hand, with nearly 13 percent unemployment versus the United States' 7 percent, and with the need for high imports, cannot solve this major social and political problem (witness the bitter coal miners' strike) by looser fiscal policies. These woulo suck in more and lower priced imports. Its only genuine solution seems to be to cut the real cost of labor (which has increased 12 percent over twenty years, while the United States has shown no such rise). Is that feasible? That is the Thatcher government's dilemma. On the whole, the United States faces a much more acceptable policy decision, if it will address it intelligently.

The experience of the French government under President Mitterand is also instructive. It started out in 1981 with a "super"-Keynesian policy of reflating the economy by large deficits and major increases in wages, and other means, only to learn the same hard lessons the United Kingdom and the United States had earlier experienced. Mitterand recently stated, "The French are beginning to understand that it's enterprise that creates wealth, that it's enterprise that creates jobs, and that it's enterprise that determines our standard of living and our place in the world. . . . You can't continue to crush with taxes and fees all those people who create wealth in France." His visit to Silicon Valley early in 1984 must have been a real eye-opener!

The West German economy has been far less expansionist than the United Kingdom's and much more solidly based on steady growth; but the absence or scarcity of entrepreneurial companies, venture capital, and new issue equities markets (for example) have hindered job creation, and in the world disinflationary environment has produced an unprecedented 9.3 percent unemployment rate. This is also true to a substantial degree in France, where large institutions also dominate the capital markets and do not encourage entrepreneurship and risk-taking. In the United Kingdom, the availability of capital for startup firms is greater, but the experience is still too recent to have had a significant effect—it may well in time. Domination of financial markets by large institutions has fortunately not yet reached the same levels in the United States.

Former Ambassador Burns has recently pointed out that the growth of the West German welfare state, and the high percentage of its GNP going to government have reduced the traditional German work habits and discipline.

The comparative figures (past and future projections) on growth rates for these three areas where free enterprise dominates (the United States, Western Europe, and the Asia/Pacific countries), are given in Tables 1–5 and 1–6. Close study would probably reveal that where the burden of government is low, the growth rates are high.

The indicators from the preceding analysis are thus unmistakable—enterprise and the entrepreneur, and what makes him or her flourish are the keys to innovation, economic growth, jobs, and rising living standards. Governments must first of all therefore seek to establish the proper climate for enterprise and entrepreneurs. Innovation cannot be forced—it can only be fostered. The total absence of entrepreneurship and the heavy bureaucratization in the Communist states underline this point also.

In Japan, where individual entrepreneurs and new businesses are rare, the entrepreneurial effort comes from within established companies, utilizing a variety of organizational patterns and greatly aided by the close support from government and the banks. However, their consensus method of decisionmaking is not especially conducive to risk-taking, particularly if rapid response to changing market conditions is necessary. Herein, therefore, lies a major American advantage if encouraged to the fullest extent. If the Japanese solve their innovative problem, America will lose the last advantage it has—where it is clearly No. 1—unless the United States makes the right decisions to raise its level of innovation and entrepreneurship markedly.

Table 1–5. Annual Real Growth Rates in GNP (%).

	1961–65	1966–70	1971–75	1976–80	1981	1982	1983
United States	4.7	3.2	2.6	3.7	2.6	−1.9	3.4
Japan	10.0	11.2	4.6	5.0	3.2	2.5	2.0
Canada	5.7	4.8	5.0	3.1	3.8	−5.0	3.8
France	5.8	5.4	4.0	3.3	0.2	1.5	0.5
West Germany	5.0	4.2	2.2	3.5	0.2	−1.2	1.2
United Kingdom	3.1	2.5	2.1	1.6	−2.0	0.5	2.5

Source: Joint Economic Committee

Table 1–6. Regional GNP Projections, 1984–2014 (1983 Exchange Rates and Prices).

	Annual Growth Rates (%)			$ Billions, U.S.		Percentage of U.S. GNP	
	Labor Force	Output/ Person	Real GNP	Year 1984	Year 2014	Year 1984	Year 2014
United States	0.9	2.6	3.5	3,500	9,900		
Western Europe	0.3	2.5	2.8	2,850	6,500	80	66
Asia/Pacific[a]	1.2	4.2	5.4	2,050	10,000	58	101

a. Asia/Pacific includes: China, Hong Kong, India, Indonesia, Japan, Malaysia, Philippines, Singapore, South Korea, and Taiwan.

Source: Texas Instruments Inc.

LARGE COMPANIES ARE ALSO CRUCIAL

Emphasizing the role of venture capital and the entrepreneur in technological innovation and job formation does not require denigration of the role of big industries or companies in maintaining the national economic health and creating wealth. A majority of the GNP and U.S. exports come from this sector, and some 20 million Americans are employed by large companies. They pay the taxes which provide the incentives and money for young companies to prosper, and governments to function. They buy from many smaller companies, and often encourage their role as suppliers and subcontractors. They maintain American competitiveness domestically against imports. Thus, the environment for stimulating large companies is equally important. Business managements, on their part, have become increasingly aware of the need for change (those who do not will soon encounter declining fortunes). In addition to the other trends noted, they have been seeking to develop internal methods of entrepreneurship—"intrapreneurship" as it has been called. Many more companies now see that technological innovation is the key to competitiveness and growth. The hard American dollar, and the difficulty small companies have in the world markets, make the redoubling of large company efforts to increase exports especially important. However, protectionism—the weak subsidizing the strong—must be resisted wherever practical even more strenuously than ever, as it breeds inflation, noncompetitiveness, and international instability. Large companies have the staff and resources—money, marketing, science, and technology to evaluate large projects and take greater risks, which small companies cannot do, and thereby also raise American productivity.

It is incredible that despite all the history recited herein, there are still voices being raised that business is getting rich at the expense of the individual. One must be very firm about the real situation: business large and small is the *producing* sector of the economy from which wealth and jobs flow; individuals outside their employment roles are *consumers.* The wealth created by business eventually goes to individuals; the costs of doing business, including taxation, are also eventually paid by individuals. Wealth cannot be generated by rearranging existing assets, or redistributing it, or plundering it (as in some mindless "takeovers" of today). It can only be created by capital and human investment in production of new products, processes, and services, as well as improving those which now exist. Companies and businesses are society's institutions to which are delegated the task of this wealth creation. The nation that does not recognize this great lesson from history is doomed to repeat its past failures.

POLICY CHOICES IN 1985

The discussion thus far could be summarized as follows: The United States is a technologically based nation. Technological innovation has been the largest factor in its economic growth, international competitiveness, and job creation, even though it also has a darker side. All progress in history has had its costs. Much self-examination is taking place within the scientific and technical community, and among American entrepreneurs and managements. What is equally required is that those now involved in the political process, and economists, must make a greater effort to take better account of the positive role of technological progress.

What is the outlook for such greater understanding in 1985 and beyond? The administration and Congress are just coming to grips with the question of further budget cuts; the outcome is quite uncertain at present. It should not be overlooked, however, that the increased defense program has the potential of much valuable commercial "dual-use" generic technology. As in agriculture, defense and the private economy have worked well together, with a good record of technology transfer (very high-speed integrated circuit (VHSIC) programs, aerospace, supercomputers, new materials, telecommunication, and more). Furthermore, if sacrifice for larger ends becomes necessary, care must be taken that economic growth, jobs, and competitiveness are given priority, nevertheless, with full realization of the pivotal role technological innovation plays.

TREASURY I

At the same time, the country is confronted with a plethora of new tax proposals—making it possible that there will be a major tax bill in 1985 or 1986 as there was in 1981, 1982, and 1984. The dangers that face the country if inappropriate tax policy is adopted can best be illustrated, in light of what has been said in this chapter, by some comments on proposals made November 1984 by the Treasury, which would form the most radical change in the tax laws since the income tax was introduced in 1913. Although the president's proposals of May 1985 are different in some respects, and the final bill, if any, that emerges from the political process will certainly be different still, it is instructive to review certain aspects of these proposals (dubbed Treasury I and II) in light of what has been said here. First, consider Treasury I, which was the point of departure for Treasury II:

1. The most fundamental philosophical question is whether the United States wishes to abandon its principal postwar policy of favoring in its tax policy certain economic and social activities, hoping that

everything will just work out in a "neutral" environment, while its competitors abroad (most especially the Japanese) follow the *opposite* strategy. It may be added that the United States does not follow a neutral policy in its budgetary approach, or in trade matters. In fact, even this proposal is not neutral. All "profits" are not the same across industries and companies, and there is no realistic way they could be. Housing is still a favored investment because interest on primary home mortgages would be deductible in full; to maintain adequate investment in productive facilities, favorable tax incentives for business must also be contained in the code, especially in industries that face severe international competition. Tax-exempt institutions still compete with private individuals for investment opportunities, often to the latter's disadvantage, as they must act on after-tax considerations. Enterprise zones have been praised by the president as a special incentive for high risk-taking but there are many other high-risk enterprises. The capital gains tax experience of 1978, and the impact of other features of the Reagan tax programs, argue strongly for continuation of proven effective policies. In short, nowhere has neutrality of tax policy as advocated by certain tax policy purists ever been applied, let alone shown to be worthwhile; hence, proponents of such a philosophy have the heavy burden of proof. There is little evidence that these proponents have seriously studied the realities of the business and technological worlds in reaching their conclusions, or that they understand the implications of modern economic theory and research. At the same time, there are important social goals embedded in the present code which should not lightly be abandoned or damaged: low-income rental housing; charitable contributions (where the United States has achieved a truly enviable mix of social benefits); state and local tax deductibility, which provides for favorable education and aid to the disadvantaged of much of the United States; tax-free accumulation of various insurance and other benefits by employers for the account of their workers, and others, all drastically altered in the Treasury I proposals.

2. Another illustration is afforded by the experiences of many companies which took advantage of the accelerated depreciation (ACRS) and investment tax credits (ITC) in the 1981 and 1982 tax bills, designed to encourage new capital investment and increase American technological competitiveness. It is not well understood that perhaps only twenty companies export over $1 billion of goods annually, among them General Motors, General Electric, Boeing,

Ford, Dupont, IBM, Kodak, and United Technologies. Many of these used such tax credits to advantage in their work, and funded investment without recourse to external debt. Particularly in an economy of high interest rates, which favors housing and where hurdle rates for investors are high compared with the Japanese, such tax features are essential to encourage new investment. Indeed, in preserving American companies' competitiveness, a higher proportion of equity capital is being employed so as to combat Japan's lower interest rates and more accommodating financial system in the face of economic uncertainty. This is done despite the higher cost of equity versus debt and underlines the need for compensating incentives for new investment. The recovery from the 1981–82 recession has featured an extraordinary investment boom. Spending on plant and equipment rose more during the first seven quarters of the current recovery than during corresponding periods of any of the last five economic expansions, being more than double. The tax policies of the 1981 and 1982 acts were certainly conducive to these results, although it is too short a time period to provide good correlations. The real question is whether such an investment surge would have been as great without these tax incentives, and this seems most improbable. Furthermore, much of this investment is in modernization and cost reduction, thus restraining inflation and promoting competitiveness by improving productivity and product quality. Half is in high-tech investment, and half of that is in computers alone! However, as noted, if the hard dollar persists, Stephen Roach (1986) of Morgan Stanley points out that U.S. capital goods producers may be increasingly squeezed out of participating in the most dramatic transformation in the U.S. economy since the Industrial Revolution. The mostly young high-technology companies have boomed and have created jobs at the cutting edge of economic growth. Yet, too many American companies are still illiquid. Here, also, the health of the total equities markets matters profoundly. This is no time to destroy the successful tax policies of the Reagan first term. The more capital intensive the firm or industry, the more significant does the tax policy affecting it become; but it is more and more clear, *all technology is capital intensive.*

One example pointed to in the recent press is the case of General Electric, which paid no income tax in the period 1981–1983 while earning \$6.5 billion domestic pre-tax profits. But, in addition to various local and payroll taxes they did pay, GE invested \$18 billion since 1981 in new plant and equipment, which may well have sustained at least 250,000 jobs per year in the United States. Does anyone seriously contend that the government could have spent the deferred (its effective rate is about 32 percent) or reduced taxes more wisely than did GE—an outstandingly well-managed company prominent in its R&D and strategic planning approach and headed by a Ph.D. engineer? In fact, if lowering the corporate tax rate requires abolition of such favorable tax treatment for investment, then there is a high probability that many companies would elect to *reduce* innovation and increase their dividends pay-out, especially considering the partial deductibility of dividends that the proposals contain. The better solution is to amend the ACRS and ITC regulations to correct known and unproductive abuses, which cause much inequality and envy between firms. These incentives to invest serve as at least a partial offset to the triple taxation of capital in the present income tax system as it pertains to shareholders. John Shoven and Toshiaki Tachibanaki in a comprehensive paper (1985) have calculated that the overall effective marginal cost of capital to Japan, taking all levels of taxation into account, was −1.5 percent at the actual inflation average of several indices, while the U.S. figure was 37.2 percent for the year 1980. Thus, in Japan, the marginal investor is *slightly subsidized.* Further calculations in their paper indicated that at the corporate level, the effective marginal tax rates at different measures of actual inflation are in the range of 7 to 17 percent, so that they have an approximation of an expenditure tax. Japan since World War II has adopted a firm strategy of favoring savings and investment over consumption by a variety of incentive policies described in their paper, and these have proved highly effective; before the war, Japanese savings and investments were much lower. This provides another example of how careful incentives do indeed work in this direction; prior to the events of 1978−1981, the United States had few such incentives. This highly uncompetitive situation needs far more tax reductions for firms and individuals that invest in new technology and equipment than even the 1981 bill has provided; the Treasury I (and II) proposals move in exactly the opposite direction, almost suicidally so!

3. Neither the Treasury nor anyone else seems to have made reliable economic forecasts for the overall impact of this bill on the economy and its component parts, especially for the longer range. The few that have emerged since November 27, 1984 are somewhat negative about the proposals in the short run, with quantitative effects probably overshadowing any favorable qualitative effects, but what the

effects would be over the five to seven years necessary for the economy to absorb them are very difficult to predict by any model, because of the great complexity of the interrelated effects. Yet it is the overall long-term result that is critical. New classical economists would say that any change in policy can alter the entire structure of an economic model, not just measure the specific effect of the policy change. How can the U.S. government seriously make the entire economy of the United States a guinea pig in this way? There was no political demand for such drastic experimentation, only a vote not to raise taxes; the performance of the Reagan first term does not require surgery, just a corrective. Engineers would be severely criticized if they built a large-scale plant without going through the pilot plant and other preliminary phases.

4. The overall impact of the proposed bill is not only anti-investment (it will raise the cost of investing in equipment by perhaps 15 percent at a time when Japanese capital costs only half of that in the United States, and reduce incentives to save)—it is *anti-risk-taking,* the precise opposite of what is needed to remain competitive and to encourage growth and job formation. The partial dividend exclusion will favor established low-growth companies; the capital gains tax provisions will militate against new companies and risk-taking; the provision for partial indexing of interest will cause a great influx of buyers into government and other fixed income securities at the expense of the equities market; indexing of depreciation for inflation will attack new investment after 1988, but not existing investment; the elimination of the progressive corporate tax on small companies will make it harder for them to finance their growth; and the nondeductibility of state and local taxes will cause a rearrangement of business and individuals who will then be too busy to gaze at distant horizons in time. These are only a few examples, but they illustrate what appears to be a strong bias in favor of safe and secure government obligations.

   Consider the case of a New York investor purchasing Treasury bills (at whatever interest rate eventually occurs) with inflation at 4 percent. Treasury I proposes to exclude 40 percent of the interest from taxation, so that the new maximum effective tax rate would be 0.60 x .35 = 0.21 (21 percent). On the other hand, an entrepreneur starting a company with stock at a nominal price, which appreciates quite substantially in five to seven years would pay nearly 35 percent plus the applicable but nondeductible state tax (nearly 14 percent in New York) as a capital gain. This tax rate of nearly 49 percent is

"Hi, Mr. Tepper. This is the I.R.S. Say, back in April, when you paid your tax, we had no idea of the sort of bills Uncle Sam would be running up, and—well, the long and the short of it is that we have to soak you again."

more than *twice* the tax rate on investments in Federal securities, not counting a possible New York City tax as well. What incentive is there for risk-taking, or entrepreneurship? The return on venture capital would have to average more than twice the return on bonds just to break even; but since the losses due to failure are not deductible, there must be a substantial risk premium as well. Hence, the expected return would have to be three or four times as much from the venture as from the bond. This is obviously very unlikely even in the most ideal circumstances, and risk would be avoided completely. The proposed increase in capital gains taxation not only flies in the face of hard recent experience—it is a signal to the entire country to beware of risk-taking. And the data show conclusively that raising this tax will reduce revenue, certainly not increase it. In any event, the amounts are too small to matter in the deficit; yet this very tiny segment of a huge total capital market is the lubricant of entrepreneurship and technological progress. The capital gains differential and stock options are at the very most only offsets to the deductibility of home mortgage interest. Why take this grave risk for some theoretical reasons of "neutrality"? Neutrality will almost certainly translate into safety and risk aversion; there must be non-neutrality if there is to be risk-taking! *Most important, these proposals are not neutral as between investment and consumption—they clearly favor the latter.*

To be sure, nontaxable institutions also invest in venture capital and new issues, but the data show the decisive importance of the tax to individuals—the entrepreneurs, the investors who lead institutional commitments, and those who buy stocks (42 million now). Stanley Pratt, publisher of the authoritative *Venture Capital Journal,* has recently pointed out that "the organized venture capital industry has currently committed to it about $16 billion, of which at least 62 percent is from taxable sources and only 38 percent from tax-exempt investors. If one considers informal venture investments from taxable individuals, which some analysts estimate to be greater than the organized activity, tax-exempt investors are even less significant. . . . Even though many like to credit them with a leadership role, the fiduciary responsibilities of pension, endowment and foundation investors mandate a follower's role in new investment trends, after others have served as pioneers." He explains that the $16 billion comes from independent private venture capital firms ($11.4 billion), small business investment companies ($1.6 billion) and corporate subsidiaries ($3.0 billion), both of the latter being almost exclusively taxable entities. He also points out that the entrepreneurs and their employees *do* perceive the importance of the capital gains differential, which is the most potent device in the code to stimulate the creation of *new* wealth, whereas the Treasury proposals for decreasing the tax rate for dividends and interest favor those already wealthy, and discourages them from further risk-taking, the same points made in this chapter just above (Pratt 1985).

Recently, the Joint Economic Committee of the U.S. Congress (1984) released a study which shows that institutional nontaxed venture capital flows more to the better known, less-risky, and larger investments and venture funds, while the taxable funds go more to the seed capital and startup companies, where the smaller venture capital companies are essential—that is, those that are the greatest risks. Thus, as can be seen by comparing Figures 1–3 and 1–5, the growth of venture capital coincides with the rise in real interest rates. Therefore, growing companies could not have been financed by borrowing but by plowing back capital gains realizations. As the capital gains tax was reduced, the stock market revived and the Initial Public Offerings (IPO) increased. It is also known that institutional capital has been less abundant in the recent months, because it sees lower returns ahead as the venture capital industry, in the face of a poor IPO market, is compelled to increase its secondary and tertiary financings. It is necessary to understand the dynamics, and not be fooled by any simple average calculations.

The entire equities markets, if adequately liquid, provide the outlet for risk-takers to "cash in," for investors to move from one asset to another, and for companies to raise capital in the face of high real interest rates. It is impossible to single out one narrow part of the equities market (such as venture capital alone) for favorable tax treatment, for all these reasons, and many more. The markets are all interrelated. Institutions are not heroic—people are! It was from considerations such as these that the people most knowledgeable about how the system really works urged the government to retain the capital gains tax differential. In February 1985, the American Business Conference, the American Electronics Association, the U.S. Chamber of Commerce, the National Association of Manufacturers, the National Association of Small Business Investment Companies, and the Securities Industry Association, among others—including high technology and basic industries, small and large businesses, mature and emerging companies, agriculture and manufacturing, and Wall Street and the venture capital community, all joined in a strongly worded statement that said, "In our view, the Treasury Department's capital gains tax proposal, if enacted into law, would reverse two of the most successful tax changes in history, the lowering of capital gains taxes in 1978 and again in 1981." They attributed this particularly to the fact that by these changes the government had restored the reward for risk-taking and investment. They also pointed out that maintaining incentives for venture capital alone does not address the fact that elimination of the capital gains differential for all other equities will raise the cost of capital to American firms at a time when Japan already enjoys at least a 2:1 advantage in this regard. All equities investors (42 million) are influenced by after-tax considerations. The words quoted earlier from Herbert Simon (1985) need to be reiterated forcibly at this point. Instead of dealing with their static models of the economy and theoretical considerations of neutrality, the Treasury should pay attention to the way individuals and firms *actually behave,* and how they adjust their actions if unfavorable tax or other policies are promoted based on such theoretical models, which the Treasury admits it still uses. Business firms use dynamic models regularly; why is the government so retrograde?

It is curious that the Treasury proposals support R&D credit for firms but deny them the tax credits needed to implement the results of R&D! The lack of neutrality of the proposals is again shown by this example favoring only one part of the innovation process. It would be more appropriate to show concern for the substantial number of excellent research organizations that have disappeared or are endangered, by the vicious takeover practices now permitted by law, and to propose appropriate corrective legislation, and correct economic policy.

Does this bias imply that the federal government under President Reagan foresees indefinitely large deficits which must be financed with such federal obligations? And if so much attention is paid to indexing against inflation at a time when inflation rates are low, are the American people to expect that this administration really considers that inflation will come roaring back again and is preparing for it? If so, it is a naked admission of failure, and the history that will be written will be very harsh. Yet it is unjustified: the Reagan administration has some real economic accomplishments to its credit and a chance to create a long-lasting favorable climate for growth.

5. Any new efforts in the direction of tax simplification are to be applauded, but great care must be exercised before major steps (or even lesser ones, if untried) are undertaken. Many economists have shown that the most fundamentally positive approach to encouraging growth and prosperity would be to move toward an overall consumption tax system. Several studies comparing the efficiency loss to the economy from distorting the consumption/investment choice relative to the decision about the types of investment to be made suggest that such intertemporal distortions are several times more important than the distortions in the allocation of investment. Thus, a consumption tax system could be truly neutral among investments and classes of investors (such as tax-exempt or taxable, residential or productive) leaving investment decisions to the marketplace, and could eliminate tax shelters, mergers, or buy-outs of an unsound or highly leveraged character, increase savings, etc. Jorgenson (1984) has shown, using an econometric model incorporating rational expectations concepts, that a pure consumption tax generating the same total revenue as under today's codes, would permit

growth in the wealth of this country almost three times more (in percentage) than a pure flat tax (something like Hall-Rabushka, or roughly Kemp-Kasten, but the Treasury I proposal would be much less effective than these proposals in the short run, and the longer term effects are difficult to predict because of its great complexity). Because no econometric model can be "correct" in real terms, Jorgenson's model should be looked at particularly to distinguish the relative advantages of alternate policies. One could approach such a system *incrementally,* for example, by broadening substantially the scope of Individual Retirement Accounts (IRAs) and similar tax-free savings, which do impose taxes when money is withdrawn (consumed), and by adding a value-added tax on certain types of goods, while lowering marginal tax rates, or by phasing in over several years full expensing of capital investment in the year it is made. Or, perhaps as *The Economist* (1985) recently suggested, by a large increase in the tax on gasoline, or a tax on imported crude oil, to compensate. What is relevant is John Shoven's (1984) observation that the tax code today is already midway between an income and consumption tax at the personal level. Why not move closer to an all-consumption tax system?

The recent book by Charls Walker and Mark Bloomfield (1984), *New Directions in Federal Tax Policy for the 1980's,* and the Brookings volume edited by Alice Rivlin (1984) discuss consumption or similar taxes in great detail. They are admittedly radical departures from precedent. Any transition from an income tax to a consumption tax would be very difficult to engineer, particularly as a way must be found to reward risk-taking—the key to innovation and growth. But, if the pain of a wholesale restructuring of the tax code is ever justified, it should be only for a very large probable benefit which a consumption tax represents, and not for an extremely uncertain outcome such as the Treasury income tax proposals, which do not even have the virtue of greater simplicity than the present code (the indexation features and the transition rules which will endure for at least ten years, guarantee that this will be the case).

6. The proposed increase in corporate taxes would add over 30 percent to the corporate sector taxes while reducing those of individuals by 8.5 percent. Corporations would pay $45 billion more in fiscal 1990. (However, those corporations that require heavy capital investment would pay much more and those that do not might pay less. As stated above, the United States cannot afford to abandon its manufacturing sector by such methods!) This is absolutely the

wrong way to go in an era of increasing need for technological innovation in a competitive world economy. In Ralph Landau and N.B. Hannay (1981) an extensive analysis of the relation between taxation, technology and economics was undertaken by a number of experts who had real world experience. It is worthwhile to quote the major policy conclusions for policymakers drawn from their papers by the editors: "You are justified in making—indeed obliged to make in view of the national urgency—extensive business-oriented and pro-investment personal tax cuts. It is better to include a very broad mix, all at once, rather than piecemeal, as a real business stimulus to investors." There was no real consensus on the desirability or extent of general personal income tax rate cuts, and no such consensus exists today. There is little doubt that lower marginal tax rates would be beneficial in the long run, but these can only be truly feasible when government deficits and hence the required taxes can be reduced.

Under conditions where greater economic growth and investment are essential, as today, it is desirable to target incentives for *new* capital investment. If the incentive is too wide, extending also to existing capital, it is less efficient in its "bang for the buck," that is, stimulus of new investment per dollar of lost revenue. Thus, an investment tax credit, which applies only to new investment, is more efficient from a revenue point of view than lower corporate taxes, because these also benefit old capital investments. Another example is the first year expensing of new asset investments, giving a zero effective tax rate on the new investments but not applying to returns from old capital. Alan Auerbach (1984) of the University of Pennsylvania has dealt extensively with this subject in a chapter of a book devoted to improving economic growth (Wachter and Wachter). There is still an unsolved problem of how to provide investment incentives for companies with tax losses that cannot take direct advantage of such incentives. Often, such companies require high equity/debt ratios because of their risky nature; low capital gains taxes tend therefore to offset the risks and attract investors and new capital. The R&D tax credit is similarly of little use to such companies, unless there is an actual rebate from the Treasury (a most unlikely occurrence), but R&D partnerships have tended to provide at least a partial offset. Thus, neutrality, as between old and new capital or between risky and safe investments, is clearly undesirable, another fault of the Treasury I proposal.

In fact, the first Reagan tax bill (ERTA, the Economic Recovery

Tax Act of 1981), even though modified somewhat later, had many of the desirable characteristics and has been quite successful to date. It targeted its incentives very well, even though some companies and activities might not have been as socially desirable as others, and some features of accelerated depreciation were of little value to rapidly growing high-technology companies which are less capital intensive than the smokestack industries. Since the cost of capital to the former companies is often based primarily on equity investment, however, the capital gains provisions have been important to them; the situation would have been even more favorable if interest rates were lower, contributing to more robust equities markets generally. It is not necessary to demolish its basic accomplishments in 1985, although improvements in specific instances may be justifiable.

More generally, the high cost of capital to American companies mandates that the budget deficits first be reduced by cutting expenditures so that interest rates fall, and equities rise. Until that happy day American entrepreneurs and industrialists need all the tax advantages that ACRS, ITC, capital gains, and the like provide them as a partial offset to the low Japanese capital cost, as well as the abundance thereof.

It must be clear from everything said before that uncertainty leads to very short-range viewpoints, and lack of long-range planning which is essential for technological progress and growth. Any sweeping change will inevitably lead to just such uncertainty, and economic regression. The very announcement of this Treasury I proposal, and of the debate in Washington, has already introduced significant delays in investment planning by individuals and companies. Yet, any fundamental overhauling of a tax code which has been in existence for many years (although with numerous modifications) *must* require extensive national deliberation and examination; the very uncertainty of its effects requires an exhaustive investigation of every feature. Hasty action for political ends will be judged very harshly indeed as future unfavorable trends unfold. *Markets hate uncertainty!*

The greatest danger facing the country is a gridlocked political process which would result in small budget cuts and a tax increase incorporating the worst features of the various tax proposals. In such a case, all semblance of tax neutrality and logic would drown in a sea of hasty political compromise! That would be a recipe for economic failure!

THE BUDGET AND TREASURY II (OR WHITE HOUSE I)

By the middle of 1985, it appeared that some reduction of the budget deficit increases would ultimately be enacted, but nothing like the initial hopes. No one can truly say that it will solve the problem, because meanwhile the slowing of the economy (as compared with 1984) has forced the government to revise sharply upward its forecast of the deficit in future years. No one really knows what the correct forecasts should be. Furthermore, the next recession will reduce revenues and increase expenses even more so that, in all probability, \$200+ billion deficits are here to stay. Is foreign capital flow likely to continue into the country under such conditions? Michael Boskin of Stanford University (see Chapter 10 of this volume) and others have shown that the poisonous effect of such deficits will have major impact by the end of the decade. What likelihood is there for further timely action? The Reagan administration seems almost to have stopped worrying about the deficit and is placing most of its emphasis on "tax reform." As a result, Congress likewise has slowed its urge to deal with the deficit, and has begun once again to talk of tax increases as a way around the problem. Yet, as the *New York Times* (1985b) says, reducing the deficit is essential, but tax reform is merely desirable. Without strong leadership from the administration, it is unlikely that there will be a firm resolution of the priorities. Time is not working to the advantage of the country. For the first time since World War II, the federal government in May 1985 spent more than twice as much money as it took in. This set a new monthly budget deficit of \$40.5 billion! At the same time, the trade deficit was the second highest on record at \$12.7 billion. Danger flags are indeed flying, but politics as usual seems to continue.

In all probability, the next administration, of whatever political composition, will be confronted by a crisis. This could even be heralded by a Democratic victory in the Senate in 1986, which would assure total "gridlock" for the last half of the second Reagan administration. Americans and their political system do not excel at taking timely preventive action; crises seem to be the only way to concentrate the mind of the body politic (the recent experiences of Argentina, Brazil, Mexico, and other countries demonstrate how difficult it is for democracies to discipline themselves before the crisis becomes intolerable). At that time, fundamental decisions may well have to be taken: either reinflate the currency to reduce the effects of the debt, or raise taxes, and struggle through a severe economic downturn. A "soft landing" will then be impossible. The growth rate of the economy in either case will slip further behind the Japanese; greater protectionism would become inevitable. Wall Street has shown a mild enthusiasm for the actions thus far and the economy is picking up somewhat, but most of the modest gains have come from an easier Federal Reserve policy. This cannot go too far, or inflation fears in the financial community will reawaken. There is no real alternative to a tighter fiscal (without tax increases) and looser monetary policy if growth is to be continued. Thus, the macroeconomic climate for innovation and technological progress seems to be becoming increasingly cloudy.

Into this uninviting prospect the administration has released Treasury II. Because it departs in a number of significant respects from Treasury I, the tax policy analysts in the Treasury would probably prefer it to be known as White House I, but that is futile. Many comparisons have been made public between these two proposals, such as that the capital gains differential has been restored, the indexation of interest has been eliminated, oil and gas incentives have been partially restored, lower taxes for small corporations have been reinstated—all reflecting the realization that Treasury I is extremely anti-risk taking and therefore antigrowth. Nevertheless, Treasury I is now ancient history (except insofar as the House of Representatives may revert to some of its worst provisions in an attempt to develop a "Democratic" tax bill, which will be even more consumption and less investment oriented). The proper comparison to be made is between Treasury II and the current tax code, which has primarily evolved since World War II. Again, this chapter will seek to examine the principles involved rather than become mired in the minutiae of so radical a change in almost the entire code. Furthermore, by the time congressional action is completed (possibly late in 1985, more likely in 1986, or possibly never) it will look quite different in detail and probably in principle. Indeed whatever principles of taxation reform were contained in Treasury I were largely abandoned by the various concessions of Treasury II to various political expediencies and industries: the final result, if achieved, will contain few principles but much politics. There is a real question whether the final result would actually be revenue-neutral, as the administration now claims. If, in the long run, it would *increase* the deficit, it would make matters doubly bad. On the other hand it could be used to disguise a substantial increase in taxes. Therefore, the questions in the following paragraphs (not necessarily in order of importance) deal solely with this comparison. Much of what has been said here about Treasury I also applies to Treasury II.

1. *Will the growth and competitiveness of the economy be improved in the longer run?* This is virtually impossible to predict with any accuracy at all from the commonly used econometric models. Such calculations as have been made suggest a reduction in growth rate in the short run. However, it is absolutely clear that, if the investment in productive equipment is reduced, the economy will suffer in competitiveness. It is precisely in the manufacturing sector, agricultural sector, and certain service sectors, where the bulk of the R&D is performed and the exports furnished, that capital intensivity is critical to maintain competitiveness, and these would be the sectors most damaged by Treasury II. Furthermore, Treasury II proposes to reduce significantly the ability of students to obtain a college education. In an increasingly technological age, where investment in human capital is essential, this is the wrong direction. Even the R&D tax credit has been made more restrictive. The Japanese are following a very different path, which even at present growth rates would mean they would overtake the United States and become the greatest industrial power in the world, in not much more than one generation. Is the United States content to follow the path of the United Kingdom and sink to second class status? Dare it? In all the publicity that consideration by the Congress of Treasury II has generated, the talk is mostly of winners and losers (that is, of redistribution), and almost never of the future growth capability and competitiveness of the U.S. economy, grounded as it must increasingly be in technological and other innovation.

Some of this failure is due to pronouncements by economists who, although they should know better, are as yet unable to grasp adequately that the United States is now in a global market; that U.S. firms in many areas face formidable competition both abroad and at home from countries which play by very different rules; that the U.S. investment rate, even at the higher recent levels is well below that of all the major economies with which the United States trades and competes; and that the econometric models used to predict the effects of the tax law changes are still based on static and closed economic systems, treating the United States largely in isolation from the rest of the world. Thus, forecasts which say the GNP will be 1.5 percent more by 1995 with this bill, as the Treasury claims, for example, are simply without adequate foundation. *Business Week,* Chase Econometrics, Wharton Econometrics, Morgan Guaranty, and others say the growth rate in GNP will be lowered in the short run, and this certainly seems probable from the considerations mentioned herein. Short-term negative effects may well be irreversible in the long run, and even if the bill were to be a net positive in ten years, because so much would have been not built and produced in the interim, the economy could not catch up. It is inconceivable that Keynesian demand stimulation by tax cuts for the lower income taxpayers will produce a more competitive and more rapidly growing economy. For that, specific supply side incentives and provisions for risk-taking are mandatory, just as they were in 1981. Yet what is proposed is essentially a 1960s' type redistribution.

When some Keynesian or other economists applaud this bill because it will tend to be more neutral among investments, so that capital can flow to its highest return uses, they simply are ignorant of the realities of the world: that foreign competitors are willing to sell products into the U.S. market at a loss, to gain permanent market share, as is occurring in memory chips right now; that their criteria for investment are often very different than those of their U.S. competitors, particularly where the cost of capital is much lower and the capital is in abundant supply (as in Japan); that the horrendous trade imbalance cannot improve if the principal exporting companies are also the ones most adversely affected by Treasury II; that the real comparative advantage of the United States lies in its ability to innovate and engage in technological entrepreneurship in both large and small companies, in high tech and low tech, which is a very *risky* undertaking and would not be practiced where alternative safer "me too" investments are available unless special tax and other legal incentives are provided (capital gains, ITC, ACRS being the principal tools for which there is experience); and that the tax code even under Treasury II would still not be neutral as between tangible and intangible capital investment (intangibles include advertising, market development, and other costs such as R&D, which can be expensed and which for many businesses are large). For all these, and many more reasons, supply-side incentives of all kinds need to be retained and indeed strengthened if the United States is to retain a balanced competitive economy with adequate job formation. Some of the comments on Treasury II seem to gloat over the fact that it basically favors high tech over smokestack industries, and thus the "winners" over the "losers." This again betrays complete ignorance of the workings of the economy, where high technology requiring large capital investments is spreading into virtually every industry, if it is to remain competitive and permit adequate economic growth. The health of the economy is even more dependent on manufacturing than before, not only for competitiveness and inflation resistance, but also because the higher productivity levels in manufacturing lead to a great impact on business cycles when manufacturing turns down.

It is strange that while even some sophisticated economists are unable to see this issue clearly, columnist David Broder in the *Washington Post* of June 10 had no difficulty at all. He lists as first among four imperatives for any tax reform bill the issue of productivity.

> We live in an increasingly competitive international economic environment. Our workers are competing for jobs, not just with each other but with the workers of Japan, Western Europe and the Third World. Our companies are competing with theirs for markets and profits. This is a fact of life; maybe the most important reality we face. No theoretical or ideological or emotional argument for 'soaking it' to this group or that can be given greater weight than that practical reality.

His other imperatives are *progressivity, principle,* and *poverty.* Data Resources, of course, understands this issue quite well, even though there are probably valid criticisms to be leveled against some aspects of their model as for most such models. Their comment on Treasury I applies as well to Treasury II:

> Explicit subsidies to corporate capital formation may appear inconsistent with the goal of removing government influence from private decision-making. But use of this support must be reviewed within the context of other features of the tax code, such as the homeowner subsidy and our national goals—*economic growth in a competitive world economy.* Once we have a "second best" world that is not freely competitive, there can be a powerful argument for government intervention to secure our future standard of living. (Emphasis added.) (Data Resources 1985.)

In July 1985, Allen Sinai, chief economist of Shearson Lehman Brothers and formerly of DRI, issued a detailed economic study (Sinai 1985). It employs a complete model of the U.S. economy and quantitative analyses of various tax changes from partial equilibrium models of business capital spending. His conclusion is stated thus:

> The changes in business taxation recommended by President Reagan . . . are desirable if there is no national goal to enhance capital formation. But increasing business capital formation and raising business savings were important goals for the economy just a few years ago. The abrupt turnaround in national objectives is a surprise and breaks the implicit "contract" of the government with the business sector. Also, there are potentially damaging effects on inflation, U.S. competitiveness, and

business after-tax profits with the new tax changes.

In September the chairman of the Council of Economic Advisers, Beryl Sprinkel, issued a report that claimed that the president's tax bill would raise GNP, but specifically stated that no econometric model had been used to arrive at such a conclusion (which was immediately challenged by many private economists) because models could not predict dynamic behavior. In light of the critical remarks in this paper, perhaps this view represents similar realization in government circles, but that does not justify essentially unsupported conclusions either. The models used by Sinai, DRI, Wharton, and others are indeed only first approximations but when they flash danger signals, it is necessary to proceed very cautiously even if one has faith that human behavior will come to the rescue. It is impossible to visualize such a happy outcome when, as Sinai says,

> Fundamentally there is no way that business taxes can be raised without reducing business spending. Removing much of the targeted tax subsidies for business capital formation and substituting a nontargeted and lesser amount of corporate profits tax reduction should reduce output, lower business equipment and plant outlays, lessen the associated borrowing, depress employment, reduce wages, and cut after-tax corporate profits in the most capital-intensive sectors and industries of the U.S. economy. On the other hand, output, employment, and after-tax profits should rise in the most labor-intensive sectors and industries. Overall, the net increase in business taxes means less production, less capital spending, less employment, and lower profits for the business sector than otherwise would be the case. The shrinkage in business sector activity, especially in goods-producing industries, will result in less production capacity and potential output or goods production and in more inflationary pressure on product prices as the U.S. economy continues to grow.[3]

2. *Are there as many incentives for savings and investment?* Definitely not. Although the few mentioned here have reduced the disparity, there are still a number of provisions favoring consumption. Yet, it is clear that a dollar of savings now will in the long run produce more than a dollar, and thus the extra value to society of a dollar of investment must be greater than the social cost of not consuming it.

The combined effect of eliminating ITC and reducing ACRS, despite a reduction in the corporate tax rate from 46 percent to 33 percent, is an additional $120 billion or so tax burden on corporations over five years. Calculations using the methodology of George Hatsopoulos and Stephen Brooks in this book indicate that the overall cost of capital would increase by 1 percent after taking the Treasury II proposals fully into account, which makes American companies even less competitive against the Japanese. Current tax policies encourage retention of corporate income, rather than distribution as dividends, generating a higher propensity to save from corporate than from personal income. It is therefore a setback to the economy that the cash flow of U.S. corporations might, in total, diminish by $222 billion over five years. The tax burden would be 24 percent higher in 1986–1990 on companies, while individuals would get a tax cut of 6 percent. However, the overwhelming number of individual taxpayers would get only a very small cut, while many companies would suffer a severe increase. Thus the winners will gain a little but the losers will surrender a great deal. In view of the probability of an economic downturn sometime before the end of 1987 this means corporate earnings and investments are bound to fall. The forecasts may in fact prove to be much more unfavorable than the current static estimates from which they derive. This sum is being transferred to individuals as tax reductions, many in the lower brackets who traditionally save very little, while corporations invest a large part of their cash flow. This, of course, is due to the triple taxation at the individual level, which the meager 10 percent dividend deductibility (if it survives) will hardly alleviate. The ACRS and ITC have been available only for domestic investment. If ITC is eliminated and ACRS is diminished, as proposed, there is a risk that some investment will shift overseas. If together with such increased capital outflow, there is a decline in foreign capital inflow, there could be upward pressure on interest rates, since currently the net capital imports are financing half of the federal deficit. These provisions reward old capital investment while weakly improving new capital investment in the longer run, a very inefficient procedure. It is strange that the Treasury uses exactly this argument to justify taking back the "windfall" depreciation gains the lower corporate tax rate would generate (not recognizing that a lower corporate tax rate by itself reduces the value of all depreciation allowances on an after-tax basis).

If state and local tax deductibility is eliminated, savings and investment will be affected. The states most affected by this major change contain the residents most likely to save and invest. Many of these will experience a tax increase; others will start thinking about moving and will have little time for longer term investments.

Although the IRA restrictions are somewhat eased, the growing

401(k) savings plans are proposed to be cut back sharply or perhaps eliminated. Taxes will be levied on some hitherto untaxed fringe benefits and buildup in insurance policy values. Employee Stock Ownership Plans (ESOPS) would be made less attractive.

It is proposed to allocate corporate interest deductions based on the ratio of foreign to domestic income. For many multinational companies which raise their debt capital more in the United States, this means a higher cost of capital. Also, the computation of foreign tax credit would be changed from an overall limitation to a per-country limitation which would sharply affect multinational corporations that have operations in both high- and low-tax countries. Tax burdens would rise, but, really, is the U.S. government trying to drive the American companies out of foreign markets?

3. *Is the present proposal as favorable to risk-taking?* This depends on the industries. Some would be more or less out of favor as a result of this bill: timber, contractors, cattle-raising, real estate, commercial building, oil and gas, some utilities, banks and insurance companies, leasing companies, airlines, manufacturing generally, mining, entertainment and recreation. It is instructive to note which industries or companies approve of provisions in the bill: broadcasters, publishers, advertising agencies, software developers, health care organizations (but not drug and hospital supply companies which would lose their Puerto Rico exemptions), cosmetic companies, the tobacco industry, food retailers and wholesalers, and some of the very profitable but not capital-intensive, high-technology companies. (Some of these may not yet realize that there are hidden traps even for them, such as cash accounting changes.) Many of these industries face no competition from abroad, take little part in the balance-of-payments issue, and therefore can be more risk-averse, are mostly labor intensive, or are so profitable that they have all the investment money they desire. What a strange antigrowth policy this bill thus represents! Investment in federal securities will still be favored, because they are exempt from state and local taxes, while equities may not be very attractive for the reasons cited above. It still looks as if the government intends to continue its domination of the credit markets.

   Even housing, which constitutes most families' principal wealth, would be unfavorably affected; values would fall, and rentals probably increase over time. The average homeowner would be most disinclined, under these circumstances, to invest in risky equities.

Contrary to belief, individuals are still important to the stock markets, and it is most undesirable to remove them from such investment activity.

4. *Is a personal tax cut at a time of huge Government deficits justifiable at all?* From economic considerations alone, the answer must be no. It was a political decision, based on an as yet unproven theory that the majority of people will work longer, harder, and more efficiently if their marginal tax rates are modestly reduced. While there may well be some basis for this as a general proposition, particularly at the middle and higher tax brackets, it has no solid quantitative evidence to substantiate it for the population as a whole. In any event, if the macroeconomic scenario presented here is at all correct, there may have to be a tax increase before long. Why go through this very short-term experiment? Although there is no proof that the general tax cuts of 1981 (Kemp-Roth) really changed most people's behavior or motivation, they did increase the deficit mightily. Certainly, the reduction of the top rate from 70 percent to 50 percent (a Democratic initiative) was helpful in reducing unproductive tax shelters, but the rest is conjectural. Tax cuts do constitute Keynesian demand stimuli, and the 1981 cuts pulled the country out of the sharp 1981–82 recession, but the price was the yawning deficit, because a politically divided government could not cut expenditures fast enough. The only real specific investment stimuli in the 1981 and 1982 acts were the ITC and ACRS credits and the further cut in the capital gains tax. Michael Boskin (1985a) has just disclosed studies that show about 25 percent of the investment boom of the last two years can be directly attributed to the first two features, which more than offset the negative effect of high interest rates. However, the effect could have been greater if these incentives had not collided with the suddenly increasing trade deficits, which permitted imports of much new equipment. Where a specific company does not benefit from incentives, its supplier of equipment (such as computers), or customers very well may. Of course, one must be careful about just what is counted in the word investment as the word includes items often really consumer durables (such as many types of computers), and methods of financing purchases of different types of assets vary, so that the effect of taxes is often unpredictable from raw data. When President Reagan leaves the White House, it is even less likely that expenditures will be cut further.

5. *Following upon these points, will unproductive tax shelters really be reduced?* In the first place, the magnitude of tax shelters has been exaggerated. As even Public Citizen Inc. says, the total amount of shelter offers was about $18 billion per year (for 1983) mostly in oil and gas, and real estate, which while larger than venture capital, is very small compared with total investment by the corporate sector, or even by comparison with the taxes paid by the highest income brackets. In any event, a marginal rate of 35 percent per se will not reduce shelters as much as hoped because many of those high-income individuals most likely to utilize them live in high-tax states, and with nondeductibility of these local and state taxes, their marginal rate will hardly be an improvement over current law. Thus, instead of 35 percent Federal marginal rates, these rates would now apply (property taxes not included, which are additional):

California	46%	Minnesota	45.5
Connecticut	48	New York City	51
Massachusetts	45.75	North Carolina	42
Detroit	43.35	Philadelphia	42.31

Congressman Kemp has made this point too, which is why his bill proposes a flat top rate of 24 percent and deductibility of property and local taxes. Under these conditions, he believes tax shelters become uninteresting. Senator Packwood has also spoken in this vein. In fact, the Kemp-Kasten bill is generally much more favorable to growth and investment than Treasury II. There are other antishelter provisions in Treasury II, particularly for real estate, which will probably be changed.

6. *Will there be less uncertainty in planning by companies and individuals for future saving and investment?* On the contrary, the transition rules alone will require years to phase in, and people's minds will be concentrated on these shorter term years (up to ten, possibly more). Furthermore, all the legal precedents of the postwar years which lie behind the current code, will be shattered, and it will take years of rulings, litigations, and decisions to create a new basis for confidence when tax-related decisions must be made by companies and individuals. It will be a largely uncharted sea. Also, by eliminating past precedents, trade-offs, rulings, and decisions, the path is made clearer for a possible increase in taxes, and then there will be few places to hide. Finally, this bill is the fourth major tax bill in the

last five years. The frequency of these changes has got to be unsettling to decisionmaking everywhere. Such past tinkering is bad enough, but this radical proposal is also a gigantic experiment upon the entire U.S. economy, with absolutely no assurance of the ultimate outcome. It is almost certain that this will not be the last tax bill for ten years (as is badly needed) but only another in the series of experiments, although a radical one. Once, another conservative party attempted to cure the economic ills of the day with tax surcharges and tariffs. In *The Wall Street Journal* Alan Reynolds quotes President Herbert Hoover, who said on May 1, 1930 to the U.S. Chamber of Commerce, "We have for the first time attempted a great economic experiment, possibly one of the greatest of our history." It is not an example to inspire more great economic experiments.

7. *Is the proposal less complex?* There is no indication that the work of lawyers, accountants, professional tax preparers, and taxpayers themselves will in any way be eliminated. Real simplifications have eluded the writers, although a few more taxpayers may be able to use the short form.

8. *Is the proposal fairer?* This has been a key argument in the political world for such drastic revision of so fundamental an economic structure as the entire tax code. Fairness, of course, is in the eye of the beholder. Some corporations have indeed utilized the present code to take advantage of tax incentives to reduce their taxes, and this will be certainly reduced by the proposal, but at an economic price. Individuals living in the sixteen high-tax states and the District of Columbia (which contain roughly 40 percent of the U.S. population) will certainly not feel it is fair to reduce their deductibility when whole life-styles and practices have been established based on this allowance. The governments of these states will feel the bill is very unfair, coming on top of the proposed elimination of revenue-sharing, and of tax-exempt municipal securities for any industrial or revenue purposes, except for general obligations. They have the larger burden of welfare, immigrants, and poor people generally. When New York tried to limit welfare payments to those who had resided in the state one year, the Supreme Court struck down the legislation. The states can only move toward greater starvation of their less fortunate if such a bill becomes law, and reduce levels of education at a time when the United States needs a greater

investment in human capital for the new technological basis of the economy. Whole regions of the country may gain while others lose, and they will not consider the bill fair, which will only exacerbate the divisiveness of the country. Property taxes are paid by virtually all taxpayers in every state who itemize (probably nearly 40 percent, which covers nearly everyone with an income above $40,000 per year). State and local taxes are about one third of all itemized deductions, up and down the tax brackets. Middle class taxpayers will not feel it is fair to reduce the top bracket more than theirs is reduced, such as in the case of two middle-income wage earner families, and there are others who feel progressivity will be unfavorably affected. Corporations that quite legitimately and in good faith took advantage of ACRS and ITC provisions of the 1981 and 1982 acts will consider it quite unfair that the bill proposes a so-called windfall tax retroactively on some of the incentive earnings provided by the previous law. If the purpose of this bill is to improve fairness overall, this has *not* been accomplished, and probably never can be. The principal justification given by politicians for the thrust of Treasury II is that individuals feel corporations and business generally are not paying their fair share of taxes, but that has always been true, because most people are ignorant of the role of business in the economy. Recent studies show that while people generally want others to pay more, there has not been any great wave of dissatisfaction with the tax code. The current talk is largely stimulated by politicians seeking advantage over the opposition, aided by a few economists who have perennially disagreed with the thrust of President Reagan's first-term policy—a truly unnatural alliance! Voters basically like the deductions they have, including especially state and local deductions, but they are ambivalent about marginal rates. Who will ultimately gain the political advantage depends on what law actually emerges, and the longer run behavior of the economy. The present code represents years of political bargaining and trade-offs which have been necessary to provide some sense of equity and fairness over time, combined with reasonable economic efficiency. No proposal delivered from a government ex cathedra can possibly substitute for such a lengthy process of adjustment, and this one will not either. Only after many years could fairness be measured, and by then there will be more tax bills. Thus, it *cannot be said that this is a reform bill at all.* It is instead a radical change in economic policy, arising from perceived political goals, which is once more a redistribution policy.

The foregoing is by no means a complete analysis of all the pros and cons. Of course, there are some good provisions in the bill: it *is* desirable to lower all marginal tax rates; consumer credit may be somewhat curtailed, although the unpredictable behavior of many individuals will likely discover ways to continue their borrowing habits; some uneconomic investments may be discouraged; interest rates may be somewhat lower if other things were equal, which is seldom the case. But the overwhelming fact is that it is a *radical experiment with unforeseeable results.* As mentioned, the Congress will now go to work. Unlike 1981 when the Democrats outbid the Republicans to lower taxes, this time they will surely go the other way and, in claiming credit for the result, will move even farther in the direction of redistribution, consumption, reduced incentives for risk-taking and investment, and complexity. The president has little room to maneuver, whereas Congress has a great deal. It is quite possible that the top rate will be above 35 percent, that local property taxes but not income taxes will be deductible, that the middle class will get a bigger tax cut, that depreciation rules may be further stiffened, that the capital gains tax will be raised, that the 10 percent dividend deduction would disappear, that oil and gas provisions will be made less generous, and the corporate tax rate raised. Little may remain of any Republican proposals, whether I or II, and no one will know the long-run effects of a bill quickly hammered out in legislative committee rooms after all hearings have been completed and new ones virtually impossible.

Thus, it is not really a question of picking winners and losers; there will be some of both. Will the economy and the country be better off in the long run? The risks entailed in this proposal, and its ultimate outcome, far outweigh the probable rewards, and this is precisely what should be avoided in tax policy. It is far better to move step-by-step in the ultimately better direction, not rocking the boat too much each time, but testing out how procompetition and prosavings and investment steps can be taken to increase the growth of the U.S. economy. Thus, for example, marginal rates alone could be cut, capital expenses depreciated in the year taken (which is a feature of the Kemp-Kasten proposal) and a consumption tax such as on gasoline imposed, which would move the economy from its present half-consumption half-income tax a little farther in the direction of the former. Such a gasoline tax alone or some equivalent consumption or energy tax (such as on imported oil, which would aid conservation of energy at a time when people are complacent again about the energy situation) could permit marginal tax rate reductions without any other complicated provisions.

"Sir, we've come to the conclusion that it's absolutely impossible to assemble a tax plan that doesn't benefit the rich."

GROWTH IS THE REAL PROBLEM

Growth (and increasing its rate) is the real problem of the 1980s and 1990s. Less than a percent difference in growth rate for the OECD countries, and for the Latin American countries, may be the difference between despair and hope for the huge army of unemployed, and for many of the governments involved. Revolution or debt default can hang by such a slender thread. It is no longer appropriate to rearrange the furniture (existing assets) on the deck of a vessel which, while clearly not the Titanic, may be running into heavier weather. What is needed is a more modern vessel with the latest technological aids to navigation, and that requires capital investment utilized by better trained people, in short, creation of new wealth. The United States is not in a free world market, and it needs a coherent strategy to compete and grow. A really good tax code is a basic ingredient of such a strategy. As Boskin (1985b) puts it, "growth is the . . . fairest and simplest way of assuring a rising living standard for our citizens. In a rapidly growing economy, opportunities for upward social and economic mobility are maximized, and the taxpayers are willing to support a humane social safety net out of their rising standard of living." One hundred and fifty years ago, the great prophet Alexis de Tocqueville, summed up the continuing dilemma as follows: "There also exists in the human heart a depraved taste for equality, which impels the weak to attempt to lower the powerful to their own level, and reduces men to prefer equality in slavery to inequality with freedom." The experience of the twentieth century demonstrates how strong the urge for freedom is, which accounts for the slow spread of democracy. At the same time, it has been proven that when prospects for growth and rising standards of living are nonexistent, when no one has very much to consume or keep, the other impulse takes over, and the masses prefer equality in misery by accepting slavery, which can only be imposed by totalitarianisms or by fundamentalist religious theocracies. Growth is the fundamental challenge of the twenty-first century.

BUILDING BRIDGES

To come to grips with such interrelated economic, technological and political issues, the United States needs urgently to build bridges between engineering schools, business schools, and political science and economic departments of universities, as well as between the practitioners of these arts, and the political leaders. This has been an era of increasing divorcement between theory and practice. Academic engineers know less and less about industry. Academic economists know very little about business and finance, which ought to be the main subjects of their researches, although they often step out of their scientists' role to indulge in value-laden political activities, not hesitant to experiment with the whole U.S. economy. There are too many MBAs who lack technical training; there are far too many lawyers and excessively costly litigiousness; there are too few engineers and scientists. Tax analysts (largely lawyers and some specialized economists) often consider it contaminating to solicit the experiences and views of those practitioners who might be most affected or involved, presumably on the ground they are biased. Perhaps they know too much! All the more urgent, therefore, is the need to educate lawyers, judges, and politicians to the strong technological and economic consequences of so many of their actions—in regulation, antitrust, trade barriers, price and wage controls, tax and monetary policies, and a host of other areas.

It is with this sense of urgency that Harvard's Kennedy School of Government organized the conference of which this book is the product. The United States can have growth with low unemployment and inflation—the principal ideal of any economy. It need not be a zero-sum society, which is a recipe for social and political turbulence. Technology is what makes for a positive sum growth. Greater growth rates in the future are possible if one considers the tremendous opportunities offered by the new technologies. In order to realize this outcome, however, people need to listen to one another more extensively, and to profit from the experiences of history and other nations.

NOTES

1. The opinions stated herein are solely those of the author.
2. The economic literature on growth shows, however, that if the feedback effects of technological change on capital investment are properly accounted for, i.e., the "induced" capital investment brought about by technological change, then the full contribution of technological progress to economic growth is substantially larger, and is indeed the principal driving force of growth. This relationship is what makes the economy a positive sum game, and reveals the strategy necessary to make it so (see Landau and Rosenberg 1986).
3. In a paper prepared in September 1985 for the Council on Capital Formation, Washington, D.C., John B. Shoven enlarges on his work with Tachibanaki (1985). He calculates the average effective marginal tax rate for new investments in the U.S. corporate sector for the tax law as it existed in 1980, and for various changes since then as follows:

	Tax Wedge (Average Effective Marginal Tax Rate), %
1980	35 (recalculated from 37.1 in earlier paper)
1981 (ERTA)	24
1982 (TEFRA)	30
1985	31
Treasury I	43.1
Treasury II	34.4
Ways & Means Staff	Greater than 40

The Japanese wedge is shown to be 7.4 in 1980 and 22 in 1985. While these are average rates and as shown herein, different companies will (and probably should) have different marginal rates, Shoven concludes:

> The fact that jumps out at anyone looking at these figures for the recent U.S. tax laws then, is that a major reduction in the tax wedge facing corporate capital was put in place in 1981 and that its effects have been eroded with each change since then. The Administration plan [Treasury II] would completely eliminate the reduction in the effective marginal tax rate on capital investments in the corporate sector . . . While we would have achieved some gains in evening the taxes across industries, we would not have improved the incentives facing investors.

Although the United States cannot adopt the Japanese methods, the fact remains that in this critical area they are much lower cost investors, and this country cannot afford to retrogress in its tax policies affecting investment incentives, rather than improving them. A fuller discussion of Japanese fiscal, monetary, tax, and other policies—all of which need to be taken together—is contained in the chapter by George Hatsopoulos and Stephen Brooks in this volume, and by Ralph Landau and George Hatsopoulos (1986).

REFERENCES

Auerbach, Alan J. 1984. "Investment, Taxation, and Growth." In *Removing Obstacles to Economic Growth,* M.L. Wachter and S.M. Wachter, editors. Philadelphia: University of Pennsylvania Press.

Boskin, Michael J. 1985a. "The Impact of the 1981–2 Investment Incentives on Business Fixed Investment," U.S. Chamber Foundation, Washington, D.C.

_____ 1985b. "The President's Tax Plan." Testimony before the Senate Committee on Finance, 99th Congress, 1st session, June 27, 1985.

Council of Economic Advisers. 1985. Report issued by Chairman Beryl Sprinkel.

Data Resources, Inc. 1985. "The DRI Study of Tax Reform." Lexington, Mass.: Data Resources, Inc.

The Economist. 1985. (June 15): 69.

Eltis, Walter. 1984. In *The Bank Credit Analyst,* 36, No. 1 (July): 31-49.

Feldstein, Martin. 1981. "Retreat from Keynesian Economics," *The Public Interest,* No. 64, Summer.

Hafer, R.W. 1983. "Examining the Recent Behavior of Inflation." Federal Reserve Bank of St. Louis *Economic Review* (August/September): 29-39.

Holland, A.S. 1984. "Real Interest Rates: What Accounts for Their Recent Rise?" Federal Reserve Bank of St. Louis *Economic Review* (December): 18-29.

Jorgenson, Dale W. 1981. "Taxation and Technical Change," in *Technology in Society.* Vol 3. *Taxation, Technology and the U.S. Economy,* Ralph Landau and N. Bruce Hannay, editors. Elmsford, N.Y.: Pergamon Press.

_____ 1986. "Microeconomics and Productivity." In *The Positive Sum Strategy: Harnessing Technology for Economic Growth,* Ralph Landau and Nathan Rosenberg, editors. Washington, D.C: The National Academy Press.

Jorgenson, Dale W., and Kun-Young Yun. 1984. "Tax Policy and Capital Allocation." Discussion Paper No. 1107. Cambridge, Mass.: Harvard Institute of Economic Research.

Klamer, Arjo. 1983. *Conversations with Economists.* Totowa, N.J.: Rowman & Allanheld.

Klein, Burton H. 1984. *Prices, Wages and Business Cycles: A Dynamic Theory.* Elmsford, N.Y.: Pergamon Press.

Kotlikoff, Laurence J. 1984. "Taxation and Savings: A Neoclassical Perspective." *Journal of Economic Literature* 22 (December): 1576-1629.

Landau, Ralph and N.B. Hannay, editors. 1981. *Taxation, Technology and the U.S. Economy,* in *Technology in Society,* Vol. 3, Elmsford, NY: Pergamon Press.

Landau, Ralph. 1982. "The Innovative Milieu." In *Managing Innovation: The Social Dimensions of Creativity, Invention and Technology,* S.B. Lundstedt and E.W. Colglazier, Jr., editors. Elmsford, N.Y.: Pergamon Press.

Landau, Ralph and George Hatsopoulos. 1986. "Capital Formation in the United States and Japan." In *The Positive Sum Strategy: Harnessing Technology for Economic Growth,* Ralph Landau and Nathan Rosenberg, editors. Washington, D.C.: The National Academy Press.

Landau, Ralph and Nathan Rosenberg, editors. 1986. *The Positive Sum Strategy: Harnessing Technology for Economic Growth,* Washington, D.C.: The National Academy Press.

Leontief, Wassily. 1985. "The Choice of Technology." *Scientific American.* (June): 37-45.

Levinson, Marc. 1985. *Dun's Business Month,* "Small Business: Myth and Reality." :30-34.

Nelson, Richard R. 1985. *High-Technology Policies: A Five-Nation Comparison.* American Enterprise Institute Studies in Economic Policy Research. Washington, D.C.: American Enterprise Institute.

New York Times. 1985a, June 7, A2. 1985b, June 16, 20E.

Norris, Floyd. 1985. "Wainright Sees a Strong Economy Next Year." *Barron's,* (July 22): 15.

Pratt, Stanley E. 1985. Letter to Secretary of the Treasury James A. Baker, dated 5 April.

Ranson, David. 1985. "Interest Rates Aren't Really High." *Fortune,* (July 8): 147.

Richman, Tom. 1983. "What America Needs is a Few Good Failures." *Inc.* (September): 63-72.

Rivlin, Alice. 1984. *Economic Choices.* Washington, D.C.: The Brookings Institution.

Roach, Stephen S. 1986. "Macrorealities of the Information Economy." In *The Positive Sum Strategy: Harnessing Technology for Economic Growth,* Ralph Landau and Nathan Rosenberg, editors. Washington, D.C.: The National Academy Press.

Shoven, John B. 1984. "Saving in the U.S. Economy." In *Removing Obstacles to Economic Growth,* M.L. Wachter and S.M. Wachter, editors. Philadelphia: University of Pennsylvania Press.

Shoven, John B., and Toshiaki Tachibanaki. 1985. "The Taxation of Income from Capital in Japan." Paper presented at Stanford Center for Economic Policy Research Conference on Government Policy Towards Industry in the United States and Japan, May 2-3.

Simon, Herbert A. 1985. "On the Behavioral and Rational Foundations of Economic Dynamics," *Journal of Economic Literature* (March):417.

Sinai, Allen. 1985. *"Tax Reform" and the U.S. Business Sector.* New York: Shearson Lehman Brothers.

U.S. Congress. Joint Economic Committee. December 28, 1984. "Venture Capital and Innovation."

Walker, C.E., and M.A. Bloomfield. 1984. *New Directions in Federal Tax Policy for the 1980s.* Cambridge, Mass.: Ballinger Publishing Company.

Weitzman, Martin L. 1984. *The Share Economy: Conquering Stagflation.* Cambridge, Mass.: Harvard University Press.

2 TAX POLICY FOR THE EIGHTIES

Edward G. Jefferson

Reflection on the performance of the U.S. economy and industry suggests that American business is underperforming as measured against the opportunities at home and abroad. The natural resources of the United States, including its scientific and engineering talents, its skilled, creative workforce, and its stable, open political system, should spell both strong growth and success in worldwide competition.

The fundamentals are encouraging. But how well are we really doing?

The U.S. economy has been making a strong recovery from the 1982 recession. The real increase of 6.8 percent in gross national product in 1984 was impressive, especially coupled with an inflation rate of only 4 percent. However, because of deficit federal spending, in the past five years the United States has had to depend on a restrictive monetary policy to reduce the serious inflation of the 1970s, and the five-year compound real growth rate has been a much less impressive 2.1 percent. Fiscal and monetary policies have pulled in opposite directions.

The lack of fiscal control is evidenced in large and growing federal budget deficits. At the same time, the 1984 trade deficit was an alarming $125 billion, more than triple the deficit of 1982. These budget and trade deficits, which are interrelated, overhang the economic future of the United States and are reaching crisis proportions.

The misalignment of currency exchange rates—specifically the abnormal strength of the U.S. dollar—is a major factor in our trade problem. The dollar's composite value, as measured by Morgan Guaranty's trade weighted index, is about 10 percent higher than a year ago and nearly 50 percent higher than in 1980, when the current pattern of appreciation started. This abnormal currency strength is in part a product of persistently high inflation-adjusted interest rates in the United States, which in turn are connected with excessive federal budget deficits.

For most countries the sharp deterioration of trade and current account balances would result in a weakening of their currencies. But foreign demand for dollars has remained strong with large net financial inflows keeping upward pressure on the dollar exchange value. Higher real interest rates resulting from our tight monetary policy have contributed to this inflow, although this differential is now moderating. This is augmented by global need for dollars as trade payments, by funds seeking safety from political or economic uncertainties abroad, and by speculation.

At all events, the abnormal strength of the dollar means that dollar-denominated costs of production in the United States have become much higher than in European countries and Japan, with profound negative effects on the trading competitiveness of American companies.

This conference on technology and economic policy focuses on a key aspect of international competitiveness: the need to adopt national policies that better encourage economic growth through technological advancement.

The factors affecting our competitiveness are woven together in a pattern of intricate complexity. No single action—by either the public or the private sector—will provide an adequate correction. Progress requires concerted and cooperative efforts on several fronts.

Industry plays a key role. Through investment of capital in research leading to new products and processes, as well as in the modernization of plant and equipment, industry strengthens its ability to compete. It must also put in place strong international marketing programs, emphasize quality and service, and make a commitment to improving productivity. Companies satisfied with the status quo will find themselves passed by as others at home and abroad invent new products, discover lower cost methods of production, provide better service to customers, and develop new markets.

The chemical industry illustrates the effect of attention to such efforts. It has effectively cut costs, shown commitment to international markets, and strongly emphasized science and engineering, including both basic and applied research. As a result, the U.S. chemical industry has remained a strong worldwide competitor, generating a substantial trade surplus throughout the 1970s and 1980s.

Since 1982, however, the chemical industry trade surplus has been eroding because of factors beyond its control, chiefly the overvalued dollar. An example will illustrate the industry's vulnerability. Last year Du Pont closed an efficient, productive fibers plant in South Carolina. Despite utilizing the latest technology to produce goods of excellent quality, the plant closed because it could not compete against imported products flooding the country—products discounted in cost and price in dollars by the abnormal strength of our currency. As recently as 1980 about one-quarter of the apparel sold in the United States was imported. Last year the figure rose to 45 percent. Textile imports increased by one-third in 1984 alone, despite an essentially flat consumption market. The main consequence of this trend has been the loss of a half million jobs in the U.S. textile industry since 1980.

The sharp appreciation of the dollar since 1980 has significantly undermined the competitive position of U.S. industry, imposing in effect a 50 percent surcharge on American goods sold abroad and a comparable subsidy of imports sold in the United States. The total trade deficit is costing us more than two million jobs, and threatens the industrial base of the country. The loss of these jobs adds substantially to the federal budget deficit. If the United States is to bring the dollar into more reasonable alignment with other currencies, it must bring the federal budget under control. Doing so will enable the Federal Reserve System to continue its recent moves to relax the money supply, thereby further reducing interest rates and helping restore more normal currency relationships.

How should we deal with the federal deficit? First we must realize that we cannot expect to grow out of the deficits. Moreover, a multiyear reduction program is needed for credibility in financial markets. The problem is severe enough that bipartisan corrective efforts are needed to get the job done.

We should seek to reduce the deficit to no more than 4 percent of GNP in fiscal year 1986, 3 percent in fiscal 1987 and 2 percent in fiscal 1988. To begin, priority should be given to controlling spending. According to recent projections this would require spending reductions of about $54 billion in 1986, $100 billion in 1987, and $125 billion in 1988.

Let's look briefly at the dimensions of the deficit problem. Table 2–1 shows revenues and expenditures as a percentage of GNP. In 1970 three major expenditure items—defense, entitlements, and net interest on the debt—came to 16.2 percent of GNP. The 3.7 percent left over after subtracting those expenditures from receipts was available for all other federal government spending.

Ten years later you can see that receipts have increased slightly as a percentage of GNP, but expenditures for entitlements have increased sharply to 10 percent of GNP. Offsetting this has been a decline in the defense area.

Now consider the Reagan administration's estimated figures for 1985 (column 3 of Table 2–1). As a percentage of GNP, receipts have declined somewhat and defense and interest expenses have increased. The total for these three major expenses alone now exceeds revenues. A deficit exists after only defense, entitlements, and interest charges have been paid. Nothing is available for the other costs of government.

Adding the other expenses reveals a deficit for all three years. Note, however, that it was the combined growth of defense, entitlements, and interest expense that drove up the deficit between 1980 and 1985. Other government expenses actually declined as a percentage of GNP during this period.

Projections from the Congressional Budget Office (the CBO) show that the trend persists through 1989. (See Table 2–2.) Receipts remain relatively stable and miscellaneous government expenses decline as a percentage of GNP. The deficit increases because spending on the combined "big three" continues to escalate.

As this analysis indicates, the greatest potential for savings lies in reducing the growth of entitlement programs and in achieving cost savings in defense.

The administration, the Senate leadership, and others have suggested ways to curb spending. Some form of spending freeze may prove to be the most politically practical approach for 1986. An across-the-board freeze would provide a reduction of only about $38 billion. Another $16 billion worth of spending reductions would be needed to reduce the fiscal 1986 deficit to about 4 percent of GNP. The first-year freeze must also be accompanied by a continued commitment to the spending discipline needed to reduce the deficit to 2 percent of GNP by 1988.

The increase in defense expenditures must be moderated in ways that do not threaten our national security. Studies made by one team of the

Table 2–1. Cost of Running the U.S. Government (as a percentage of GNP).

	1970	1980	1985 Est.
Receipts	19.9%	20.1%	19.2%
Defense	8.4	5.2	6.7
Entitlements	6.3	10.0	10.0
Net Interest	1.5	2.0	3.3
Total	16.2	17.2	20.0
Left over	3.7	2.9	−0.8
All other expenses	3.8	5.2	3.4

Table 2–2. Selected CBO Budget Projections (as a percentage of GNP).

	FY 1985	1986	1987	1988	1989
Receipts	19.1%	19.1%	19.2%	19.4%	19.4%
Defense	6.7	6.9	7.1	7.4	7.6
Entitlements	10.6	10.5	10.4	10.3	10.3
Net Interest	3.4	3.5	3.7	3.9	4.0
Total	20.7	20.9	21.2	21.6	21.9
Left over	−1.6	−1.8	−2.0	−2.2	−2.5
All other expenses	2.9	2.8	2.7	2.6	2.4

Grace Commission suggest that if Congress and the administration will address this challenge together much could be achieved. Nothing can be accomplished if deficit reduction becomes a political football. The problem cries out for a bipartisan solution. We should recognize that further weakening of the American industrial base would itself have a negative impact on U.S. military preparedness.

Entitlement programs absorb over 50 percent of federal revenue. Social Security, Medicare, as well as civil service and military pensions are among the fastest growing elements of the budget and their growth must be controlled if the deficit crisis is to be resolved. While means-tested programs for the poor should not be cut, economist Martin Feldstein and others have recommended limiting the growth of Social Security and other non-means-tested entitlements to three percentage points below the full Consumer Price Index. On the revenue side, this could be accompanied by a similar limitation on personal tax bracket indexing.

If expenditures cannot be reduced to the extent needed to bring the deficit down to the levels I have suggested, some revenue increases will have to be considered. In evaluating tax proposals affecting business, we should begin with a fundamental question: What tax structure would best encourage stable, long-term economic growth, foster job creation through industrial modernization and growth, and enhance the worldwide competitiveness of U.S. industries? Tax proposals should be measured against four criteria.

1. The tax system should bring into better balance incentives for savings and investment with those for consumption.
2. Capital formation essential to productivity growth and global competitiveness should be encouraged.
3. Private investment in research and development and technological innovation should be stimulated.
4. Changes in the tax system should enhance the ability of U.S. companies to compete effectively in global markets, while avoiding adverse effects on the international price competitiveness of U.S. products and services.

Let me expand on each of these principles, beginning with the need for a better balance between savings and investment and consumption.

Despite some improvement, the personal savings rate in the United States remains among the lowest of the major industrialized countries—about one-half the rate in West Germany and only one-third the rate in Japan. The federal income tax system is a major contributor to this problem because it has traditionally encouraged consumption to the detriment of savings and investment. In the first place, the United States relies more heavily on income taxes than do many other countries. This country offers deductions for interest payments on loans associated with consumer purchases. And this is one of the few countries to maintain a policy of double taxation of corporate profits—taxing both the company and its shareholders.

Some economists and political leaders have recently suggested a transactional tax, such as a value-added tax, to obtain a better balance. Such a tax would help improve the national savings rate by shifting some of the tax burden from income to consumption. The regressive aspects of a value-added tax could be corrected by using some of the proceeds for relief of low-income individuals.

To consider whether tax proposals encourage capital formation essential to productivity growth and global competitiveness, examine some comparisons with one of our most important trading partners: Japan.

Until the adoption of the Accelerated Cost Recovery System (ACRS) in 1981, user costs of capital in Japan were significantly less than in the United States. Japan's advantage derived from a number of factors, including the higher rate of personal savings in Japan. The relatively low costs of capital in Japan have encouraged capital formation and investments in productivity and growth. On a gross basis, for example, Japan invests about 28 percent of its GNP in fixed capital, compared to 17 percent for the United States.

On the positive side, the adoption of ACRS in conjunction with the Investment Tax Credit (ITC) has substantially narrowed the user cost of capital gap between the United States and Japan. A study last year by Data Resources concluded that ACRS and ITC helped to prevent U.S. business capital outlays from declining as much as might have been expected during the last recession, and have since contributed to the strong recovery. Barring major changes in U.S. banking laws or increased government intervention in credit markets, we will need to continue to depend on our tax system to encourage competitive investment in modernization and growth.

If a key reason for reducing budget deficits is to enhance the competitiveness of domestic industries, it is illogical to impair capital formation in the process. It is disappointing to note that the Treasury's November 1984 tax proposals call for the elimination of ACRS and ITC, both of which have resulted in stronger business investment. Although the Treasury points out that its plan includes a reduction in the corporate tax rate, a study of the impact of tax proposals on capital investment by economist Allen Sinai found that ACRS and ITC give a much greater "bang for the buck" than a reduction in the corporate tax rate. Sinai found that ITC and ACRS, because of their greater effect on the more capital-intensive operations, provided 75 to 80 cents of business fixed investment for every dollar of corporate tax revenue lost.

The comparable figure for increased investment for every dollar of tax lost through corporate tax rate reduction is only 19 cents.

A third criterion against which tax policy should be measured is the need to stimulate private investment in research and development and in technological innovation. Much of this conference focuses on the critical role of research and technological innovation as the driving force for both economic growth and global competitiveness. Tax policy plays an important role in determining the level of private investment in both basic and applied research and in commercial development of new products and processes.

That fact will be highlighted in 1985 when Congress decides whether or not to extend the R&D tax credit that will otherwise expire on December 31, 1985. While there may be room for improvement in the existing law, a strong R&D incentive along the lines of the 1984 Senate bill is imperative. It is noteworthy that Treasury's November 1984 tax reform proposals endorse continuation of an R&D tax credit.

Recent experience indicates the allocation of investment funds between R&D and other projects is affected by bottom-line costs. Firms with large R&D budgets—those most sensitive to R&D cost considerations—have testified that the R&D credit has been a significant motivating factor to increase their R&D expenditures. The facts bear out this testimony. Congress passed the R&D credit in 1981 in response to a steady and substantial decline in R&D spending as a percentage of GNP during the 1970s. That trend has now been reversed, and the General Accounting Office notes that "available statistics indicate there has been a substantial increase in expenditures for research and experimentation since the credit went into effect." Moreover, it is important to note that in the recent recession R&D actually increased, a reversal of U.S. experience during economic downturns since World War II.

The chemical industry's relative success in retaining worldwide competitiveness, mentioned earlier, had as a key strategic element intensive emphasis on R&D. In the United States, the chemical industry accounts for only about 9 percent of manufacturing shipments, but it does about 15 percent of all industrial R&D in this country and about 35 percent of all basic research funded by industry. The benefits of this research and innovation flow not only to chemical companies, but to the entire economy. Incentives for R&D should be preserved.

Finally, changes in the tax system should enhance the ability of American companies to compete effectively in world markets. Tax provisions which adversely affect the price competitiveness of American products and services should be avoided. Let me give you one example in each category.

A recent study of border taxes turned up some interesting findings. It concluded that the greater use of transactional taxes, which under the rules of the General Agreement on Tariffs and Trade (GATT) may be rebated at the border, gives some countries a cost advantage in foreign trade. Four major European competitors, for example, enjoyed a cost advantage of better than 4 percent. In examining tax options we should keep in mind this procompetitiveness feature of value-added and other transactional taxes.

On the other hand, we should be aware of the negative impact on competitiveness that may be attached to other proposals. For example, a tax has been advocated on imported crude oil, a tax reasoned to be relatively painless in a period of declining crude oil prices. While appealing at first glance, a tax on imported crude oil would seriously impair the worldwide cost competitiveness of many domestic industries by forcing them to use energy and petroleum-based raw materials at prices above world levels. The advantages seen in energy taxes might well be realized, however, through an increased tax on gasoline at the pump. This would not have a negative impact on our industrial competitiveness since, unlike a tax on crude oil, it would be borne directly by the consumer.

The four criteria I have proposed reflect a common theme: Each seeks to bring tax policy into closer harmony with worldwide competitiveness goals. Future U.S. economic growth—and the nation's ability to provide job opportunities and sustain appropriate social and defense programs—depends on the ability to compete more effectively in both domestic and world markets.

Recent proposals to reform and simplify the U.S. tax system are creative and merit serious consideration. Reform of the tax system is needed and supported by the public and the business community should participate in its development. However, at the moment the most important objective must be a reduction of the federal budget deficit sufficient to permit increased growth in the money supply and in turn lower interest rates and an easing in the abnormal strength of the dollar. If the United States fails to address its budget deficit soon, both our trade position and our industrial base will erode further and interest payments on the debt will rise and make remedial action even more difficult. In effect, we Americans will find ourselves in a kind of debtor's prison, with our freedom of action severely restricted.

The realities of our situation have become clear. Our international competitiveness is increasingly compromised by the many effects of fiscal mismanagement, and we will not ease our trade problems until we solve our budget problem. Maneuverability on the budget is very limited. If we are to achieve spending control, we cannot put any major expenditures out of bounds, including entitlements and defense. And if additional revenues are required, we must raise them in ways that do not further impair the worldwide competitive strength of American industries.

There is no "spoonful of sugar to help the medicine go down," but strong medicine is needed and now is the time to take it. Concerted action to address federal budget deficits, and the trade deficit, can put us on the path to sustained economic health and restore the worldwide competitive vitality of U.S. industries.

3 FEDERAL TAX POLICY AND THE BUDGET

Emil M. Sunley

Two conditions have combined to make federal tax policy one of the premier issues before the current Congress. Tax revenues are no longer sufficient to support federal expenditures and the federal budget is consequently running a record deficit. At the same time, taxpayers, practitioners, and academics are increasingly dissatisfied with the income tax, the primary source of revenue in the existing tax system.

The federal deficit currently rests at over $200 billion a year or about 5 percent of gross national product. Although previous Congresses have attempted to check the widening gap, without further action the deficit is projected to continue to account for an unprecedented share of GNP of between 4.5 and 5.0 percent.

Protracted deficits of this magnitude are economically harmful. While the deficit need not be eliminated in the next several years, it should be reduced. It should not, however, be reduced solely through reductions in spending. To do so would require intolerably deep cuts in national defense and Social Security. These programs, together with net interest payments, comprised more than 58 percent of the federal budget in fiscal 1983.

That additional tax revenues are needed is clear. The more important issue is how the federal government can or should generate these revenues. Can or should the existing tax structure, dominated by the income tax, be used? Or is an additional revenue source required?

Pivotal to the answer to this question is the current state of the income tax. The income tax is burdened from the start with some fundamental problems. For example, taxing net income double-taxes savings, thus encouraging taxpayers to consume rather than save. Furthermore, in practice, it is difficult to structure an income tax that adequately deals with capital recovery, inflation, and integration of the corporate and personal income taxes.

Besides these problems, there is another fundamental one: the current U.S. income tax is, by most standards, not a good one. The tax base is badly perforated and the marginal tax rates are high. Together these two factors make taxes an important factor in economic decision-making and, as a result, the allocation of resources in the economy is distorted.

To solve this basic problem, we have two options: abandon the income tax for a consumption tax, thereby alleviating the inherent bias against savings and the problems of depreciation, inflation and integration. Or: reform the income tax and retain it as the primary source of revenue. Under both of these scenarios, if the structure as changed still cannot produce adequate revenues, additional taxes—such as a value-added tax, or new excise taxes—can be adopted.

The Ninety-eighth Congress produced several income tax reform proposals, including the broad-based, progressive-low-rate Bradley-Gephardt and Kemp-Kasten proposals, and the broad-based, flat-rate Siljander proposal. Most of the reform plans are designed to be revenue neutral, keeping separate the issues of tax reform and deficit reduction.

In his 1984 State of the Union address, President Reagan asked the Treasury Department to perform its own exploration of a range of reform alternatives, from repairing the income tax to replacing it. In conclusion to its study, the Treasury submitted its recommendation for a "modified flat tax" on income.

By rejecting income tax replacement options, the Treasury study has narrowed the more realistic tax policy debate to that of income tax restructuring. Because of its comprehensiveness, the study provides a good focus for income tax reform discussions.

The Treasury proposal takes a base-broadening, rate-reducing approach to tax reform. Under the plan, many of the existing deductions, exemptions, and credits are eliminated and tax rates are lowered. Instead of having fourteen rates ranging from 11 percent to 50 percent as under the current individual income tax, the Treasury plan has three rates of 15 percent, 25 percent, and 35 percent. The corporate income tax rate is reduced from a top rate of 46 percent to a flat rate of 33 percent.

Also like Bradley-Gephardt and Kemp-Kasten, the Treasury claims revenue-neutrality. Though total taxes would not increase under the plan, the tax burden would shift away from individuals toward corporations. Individual income taxes would on average decrease about 8 percent, while corporate income tax payments should on average increase about 30 percent.

The plan also includes several innovative features. For example, depreciation, capital gains, and inventories are all indexed for inflation. To adjust debt for inflation, only a portion of interest expense is deductible and only a portion of interest income is taxable. To relieve the double taxation of dividends, corporations are granted a partial deduction for dividends paid.

Probably none of the reform proposals will be considered as all-or-nothing propositions. The case is strong for moving to a tax with a broader base and lower marginal rates. But, the move from where we are to where we might like to be requires the repeal of many entrenched tax preferences, each of which has strong advocates. In challenging these preferences, hard questions will have to be addressed: Is the original purpose of the preference still valid? Is the tax system the most efficient means of delivering the subsidy? Should the subsidy be delivered through a direct spending program?

Particularly in the context of revenue shortages, an incremental reform of the income tax may thus be a more palatable approach. The Treasury proposal, for example, could serve as a menu from which to choose such incremental changes. Of course, the coherence and consistency of the Treasury plan could be lost if only parts of it are adopted.

THE BUSINESS SIDE OF THE TREASURY PLAN

The business portion of the Treasury plan consists of five elements: rate cuts, revision of capital recovery rules, indexing for inflation, a dividends-paid deduction, and structural reform for particular industries.

The first four elements, taken by themselves, are nearly revenue neutral. As Table 3–1 shows, the first four elements should reduce corporate taxes over the five-year period, fiscal 1986 through 1990, by $9 billion, or less than 2 percent of expected corporate tax payments. The corporate income tax increase comes instead from the fifth element—elimination of industry-specific tax preferences.

Rate Cuts

The U.S. Treasury proposes to replace the current corporate schedule, which has rates of 11 to 46 percent, with a flat rate of 33 percent. The corporate rate would be set just below the 35 percent top marginal tax rate proposal for individuals.

In setting corporate and individual tax rates, the Treasury followed a good tax policy rule; namely, the top individual income tax rate should not be too different from the top corporate tax rate. The Treasury could have proposed a lower rate for corporations (and higher marginal tax rates for individuals), so that there would not be a shift in tax burdens toward corporations and away from individuals. Had the Treasury elected this approach, however, the corporate tax rate would have been significantly lower than the top individual rates. This imbalance would prompt corporations to accumulate earnings to avoid taxation of dividends at the high individual rates. In turn, the Internal Revenue Service would have to increase its enforcement of the accumulated earnings tax. Similarly, an individual tax rate that is much higher than the corporate rate creates undue pressure on the use of personal holding companies.

Capital Recovery

The U.S. Treasury would replace the investment tax credit and ACRS depreciation with indexed economic depreciation. This change would increase the effective tax rate on investments in new machinery and equipment.

Under current law, the investment tax credit and ACRS depreciation are about equivalent to immediate expensing; that is, the present value of the tax savings from the investment tax credit and ACRS are about equal to the tax savings that would be achieved if the equipment were fully written off when placed in service. In economic terms, the

Table 3–1. Treasury Tax Proposal Revenue Effect on Corporations ($ Millions).

	1986	1987	1988	1989	1990	Total
Structural changes						
Corporate income tax rate reduction[a]	(10,920)	(29,377)	(38,530)	(43,246)	(50,525)	(172,598)
Dividend paid deduction	–	(9,803)	(20,678)	(28,983)	(38,238)	(97,702)
ITC and depreciation[b]	18,982	42,452	62,332	81,270	99,920	304,956
Indexing for inflation[c]	(3,594)	(7,264)	(11,400)	(10,990)	(10,177)	(43,425)
Total structural changes[d]	4,468	(3,992)	(8,276)	(1,949)	980	(8,769)
Other corporate changes	17,696	34,574	37,617	40,011	43,740	173,638
Total	22,164	30,582	29,341	38,062	44,720	164,869

a. Includes effect of reduction of rate to 33 percent and repeal of graduated rate structure.

b. Includes repeal of the investment tax credit, change to inflation-adjusted RCRS and repeal of scheduled increase in depreciable property that may be expensed.

c. Includes provisions for indexing capital gains, FIFO inventories, and net interest income or expense.

d. The total involves some double counting inasmuch as Treasury estimates the revenue loss from the rate cuts assuming all the base broadening is enacted. This double counting is offset in part by Treasury's assumption of very high investment levels and low inflation rates in estimating the revenue gain from ITC/depreciation changes.

investment credit and ACRS result in a zero effective tax rate on returns to new investments in machinery and equipment. Accordingly, an investment in machinery and equipment which yields 12 to 15 percent before tax will yield about the same return after tax.

Under the Treasury proposal, annual tax depreciation would be limited to economic depreciation—the actual annual decline in value of the machinery and equipment or structure. The effective tax rate under this system is the same as the nominal corporate tax rate, here 33 percent.

Though much important work has been done in attempting to measure economic depreciation, there is clearly a range of uncertainty as to the actual rates of economic depreciation for different types of assets and for different industries. To the extent that tax depreciation differs from true economic depreciation, effective tax rates on new investments will vary.

The new depreciation system proposed by the Treasury would increase the effective tax rate on new investments in depreciable property. At the same time, the proposed reduction in the corporate tax rate would lower the tax burden on the income from existing machinery and equipment. This ironic result probably is a cost of transition that the Treasury elected to tolerate in its pursuit of a system that treats alternative investments more nearly alike.

Indexing

The indexing of depreciation, capital gains, inventories and interest is one of the most novel features of the Treasury plan. The Treasury plan here goes beyond the Bradley-Gephardt or the Kemp-Kasten proposals which, except for capital gains under the Kemp-Kasten proposal, did not deal with inflation.

Indexing involves two analytically separable issues: the definition of income and the proper tax treatment of income, once defined. One could adjust the nominal dollar amounts in the Internal Revenue Code without adjusting the measurement of income, or vice versa.

Indeed, beginning in 1985, the major fixed dollar amounts in the income tax statute are adjusted annually for inflation. Congress has not provided explicit inflation adjustments for the measurement of income. Under the Treasury plan, however, depreciation, inventories, capital gains, and debt would be indexed for inflation.

While indexing, by itself, would add complexity to the tax law, ideally the gains from making the tax system "neutral" with respect to inflation are greater than the cost of the added complexity. The indexing rules under the Treasury plan are not all that complex.

Under the plan, the adjusted basis of depreciable assets would be increased each year by the inflation rate. The fixed depreciation rate would then be applied to the resulting adjusted basis. For property other than inventory assets or debt instruments, a taxpayer's original cost basis would be indexed for inflation during the period he or she holds the property.

For inventories, taxpayers would be permitted to use an indexed FIFO (first-in, first-out) method in addition to the current FIFO and LIFO (last-in, first-out) methods of accounting. An indexed FIFO would be a simpler means of adjusting for inflation than the LIFO method. Under indexed FIFO, an adjustment would be made to inventory to reflect the change in the general price level from the time an item was put into inventory until the time it was removed and sold. Such a system, while eliminating purely "inflationary" profits, would still include in the tax base real gains and losses from changes in relative values of inventories.

Economists, accountants, and members of the business community do not currently agree on just how an adjustment for debt should be made. Some have argued that the interest deduction and interest income should be reduced by the amount that is attributable to inflation: the "inflation premium." Of course, this would require an estimate of how much of the current nominal rate of interest is "real" and how much is the inflation premium. Others have suggested that the nominal interest should be fully taxed and deducted, but at the time debt is paid off, a gain or loss should be recognized to the extent that the debt is paid with deflated dollars.

The Treasury proposes a short-cut method for indexing debt. Under its plan, a stated portion of interest income would be excluded and a portion of the interest deduction would be disallowed. The fractional exclusion rate would be adjusted annually to reflect changes in the rate of inflation. Homeowners, however, would continue to fully deduct mortgage interest on a principal residence, retaining a significant tax benefit for homeownership.

Treasury's short-cut method is clearly simpler than annually adjusting principal for inflation. It can, however, overstate the inflation adjustment, particularly where a taxpayer such as a financial institution is highly leveraged.

Dividends-Paid Deductions

One of the most significant trends in tax policy in recent years has been the movement in national tax structures from classical systems of separate taxation of corporations and their shareholders toward some form of integration of corporate and shareholder taxation with respect to distributed corporate profits. This trend has grown out of an increasing concern over the impact of the double tax burden which the classical system places on income from capital in the corporate sector.

The double taxation of corporate profits distorts the allocation of capital between the corporate and noncorporate sectors, encourages corporations to retain their earnings, and distorts the choice between debt and equity finance. Integration would reduce these distortions and would thereby increase economic efficiency. It is not clear, however, that integration would increase the level of investment. The major concern is that corporate integration would reduce the "tax" on distributions and thereby encourage additional distributions and reduce business savings. Unless increased savings of individuals would offset the reduction in business savings, total savings would be reduced.

Opposition to past proposals for integration was due in part to the fact that the benefits of integration would be spread very unevenly across industries and across firms within an industry. For example, integration would provide little or no benefit to small closely held companies or rapidly growing firms which currently pay little in dividends.

A variety of methods exist for integrating the corporate and individual income taxes. One method is simply to treat the corporation as a partnership, much as is done under Subchapter S rules. That is, all the corporation's earnings, whether distributed or not, would be imputed to the shareholders. The corporate tax could remain in place as a withholding tax, which would be creditable against the shareholder's tax liability. The technical problems of partnership treatment have not been fully worked out, probably part of the reason that the Treasury rejected this purest approach.

A less complete or partial method would be to integrate the corporate and individual income taxes with respect only to dividends. This can be done by either a dividend deduction method or by an imputation or shareholder credit system. Under the former, relief is provided at the corporate level by taxing dividends at a reduced or zero rate, as the Treasury has elected to do. The imputation system, in contrast, provides relief at the shareholder level by allowing a credit for part or all of corporate taxes paid on distributed earnings. Though these two methods appear to be quite different, they can be made equivalent.

The Treasury proposes to deal with the double taxation of dividends by permitting a company to deduct half of dividends paid. This proposal would reduce the effective tax rate to 20 percent if a company distributes all of the earnings it can, retaining sufficient earnings to pay the tax. To see this, assume a company has income of $100 and distributes $80 in dividends. With a 50 percent dividends-paid deduction, the company's taxable income would be $60 and its tax at a 33 percent rate would be about $20. Thus a company distributing dividends of $80 would have sufficient income to pay the tax of $20.

Conclusion

The depreciation, indexing, dividends-paid deduction, and rate cuts together would raise the effective tax rate on incomes from new depreciable capital to between 20 and 33 percent, depending on the dividend pay-out rate. The 33 percent top rate would apply to the income from all other corporate capital including land, inventories and know-how. Companies whose capital stock is not concentrated in depreciable capital would thus have a net tax reduction. In contrast, those whose capital stock is concentrated in depreciable property or who benefit from the special tax provisions that the Treasury proposes to eliminate would tend to have a tax increase.

TAX POLICY FOR HIGH TECHNOLOGY

The Ad Hoc Electronics Tax Group

In January 1984, the Ad Hoc Electronics Tax Group, to which I served as a consultant, released a report on suggested tax policies for the decade. The twenty medium and large-size electronics and computer firms that the group comprised recognized that the income tax evolved during a period dominated by industries fundamentally different from their own. Like other reform discussions, the group attempted to separate its review from the immediate problems of deficit reduction. It took a reasonably long-term, comprehensive view of some of the major elements of tax reform in relation to high-technology industries.

In its report, the group identified five characteristics which distinguish its industry from traditional manufacturing. Citing their comparatively high effective rates, the group argued that these characteristics bias the existing income tax against the industry, and it offered recommendations for reform.

First, these companies compete in a very dynamic market. Their products tend to have shorter market lives, being made obsolete by new and technologically improved products. Consequently, they tend to have short-lived manufacturing equipment and they must invest significant portions of their capital in research and development. Accordingly, the group found that they do not benefit as much as traditional manufacturers from the accelerated cost recovery system. They also strongly supported extension of and definitional changes to the tax credit for R&D.

Second, invariably the sales and operations of companies in the industry are international in scope. The high cost of research and product development pushes these companies to the broadest possible market. On that basis, the group recommended changes in the taxation of foreign affiliates and supported the adoption of the Foreign Sales Corporation (FSC) export-incentive legislation.

Third, companies in the technology industry have high and continuing capital needs out of which they finance their large R&D budgets and their reinvestment in plant and equipment. Because the overall level of risk in the industry is relatively high, they tend to rely on stock rather than debt as a means of meeting their capital needs. The group decided that before making specific recommendations it should further study the indexing of capital gains and proposals for at least partial integration of the corporate and personal income taxes.

The last two characteristics—the industry's reliance on the creative skills and high motivation of its employees, and the proliferation of new companies—supported recommendations and study of the taxation of stock options, employee benefits, and capital gains.

The Ad Hoc Group's recommendations differ in several cases from those made by the Treasury. In the foreign area, for example, the Ad Hoc Group's report recommended increasing a company's flexibility in shifting income, tax-free, among their subsidiaries operating abroad. The Treasury, in contrast, would impose a country-by-country limitation of the use of foreign tax credits, decreasing income-shifting flexibility.

Nevertheless, companies characterized by the five elements described are considered the most likely to benefit from, or at least the least likely to be hurt by, the Treasury plan. In short, the industry sector that is least subsidized by the existing income tax preferences is the sector that will be less harmed by their removal.

The R&D Tax Credit

One of the few tax preferences spared under the Treasury proposal is the income tax credit for expenditures incurred in R&D. The Treasury would narrow the definition of qualified expenses and extend the credit, now scheduled to expire at the end of 1985, to expenses incurred prior to December 31, 1988.

Originally enacted in 1981, the credit for R&D reduces federal income tax liability by 25 percent of the increase in qualifying R&D expenditures over average expenditures made in the prior three years.

The R&D credit is incremental; that is, it is tied to *increasing* research expenditures rather than total expenditures made in a year. This design—a high credit rate on an incremental base—was determined to be the most efficient means of encouraging increasing R&D. The more research a company does relative to what it has done, the more subsidy it receives. At the same time, however, this incremental feature reduces the effective rate of the tax credit. In fact, it is closer to 5 percent, substantially less than the stated rate of 25 percent. In any given year R&D expenditures at a higher level than previous years generate a credit that year. However, higher expenditures increase the average base against which the "increase" in expenditures is measured in future years. This average base effect reduces future year credits.

The Ad Hoc Group recommended that consideration be given to lagging the average base period to, for example, the third, fourth, and fifth years preceding the credit year. By doing so, an increase in expenditures in any year would not affect the base for two years. This would increase the effective rate of credit. Of course, this would also increase the revenue cost of the credit, and in its proposal the Treasury made no recommendations to modify the credit in this way.

Also because the credit is incremental, special rules are required to handle corporate acquisitions and dispositions. In addition, the credit must be restricted to regular corporations. Individuals with no R&D expenditure history could form R&D partnerships to generate maximum credits.

Another commonly faulted aspect of the credit is that only expenditures incurred "in carrying on" a trade or business of the taxpayer are eligible. Accordingly, expenditures incurred by new companies in the start-up phase do not generate a credit. Until a company actually produces a product it is not considered to be actively carrying on its business. This restriction puts new firms at an immediate disadvantage against the larger, established firms when it comes to innovation.

The Treasury plan adopts the definition of qualifying expenditures that the Senate included in its initial considerations of the Deficit Reduction Act. While this definition is in most respects more narrow, it does redress the "in carrying on" restriction of current law.

Capital Gains

The tax treatment of gains from the sale of corporate stock has always been important to the high-technology industries because of their heavy reliance on equity finance. In fact, electronics companies were a strong influence in the passage of the 1978 legislation to reduce the tax rates on capital gains.

In its report, however, the Ad Hoc Group elected not to make further capital gains rate reductions a high priority. The report noted that venture capital, as opposed to stocks of established companies, was probably the most sensitive to changes in capital gains taxation and there did not appear to be a significant shortage of available venture capital. Citing also the political difficulties of pushing for further reduction in (or elimination of) the taxation of gains on capital stock, the group only suggested topics for future study.

The group's position would probably be modified in light of the Treasury proposal for indexation and full taxation of capital gains. Under current law the capital gains exclusion benefits stocks with high nominal gains whether those gains are primarily real or primarily inflationary. Under the Treasury plan, these two elements of nominal gain are separated. Gains which are attributable to inflation will not be taxed at all. Gains which reflect a real appreciation in value, however, will be taxed in full. Inasmuch as the stocks of successful venture capital investments tend to experience high real gains, they would tend to be more heavily taxed under the Treasury plan relative to current law.

CONCLUSION

Given the need to confront the federal budget deficit, and the unlikely and undesirable prospect that this will be accomplished by reducing spending alone, taxes should be increased. The Treasury reform proposal, though revenue-neutral in total, offers a varied menu from which to generate these revenues. Inasmuch as Treasury preserves only four of the big preferences—the R&D credit, the Foreign Sales Corporation export incentive; pension contributions and earnings, and mortgage interest—it appears that nothing will be sacrosanct in the upcoming debate.

Nevertheless, in this environment of reform, the proposed extension of the R&D credit, the retention of FSC rules, and the repeal of ACRS and the investment credit should be of major significance to the high technology industries and their investors.

4 CORPORATE INCOME TAX REFORM

James W. Wetzler

The corporate income tax plays a unique role in the present tax system, and proposals to change it will be central to any serious debate about tax reform. Economists generally disdain the tax, believing that the distortions it creates are disproportionate to its revenue yield, but the general public strongly supports it. The result of the tension between the public's desire to tax corporations and Congress's desire to ameliorate some of the adverse economic effects of the tax is a corporate tax structure that satisfies neither the preferences of the public, which still believes corporations to be undertaxed, nor those of economists. Furthermore, prospects for simplification of the individual income tax depend on corresponding changes being made in the corporate income tax, where special interests are entrenched, there is considerable misunderstanding, and the issues are highly politicized.

The debate about corporate income taxation may be dramatically altered by the Treasury Department's November 1984 tax reform proposal, which recommended significant structural reforms to the tax and an increase of about one-third in the overall tax burden imposed on the corporate sector. The surprising willingness of the Reagan administration's Treasury to recommend such a large redistribution of the tax burden to corporations appears to confirm long-standing claims that the corporate sector can afford to pay higher taxes without adverse impacts on economic growth, and can be expected to influence the debate about corporate taxation for years to come regardless of the fate of the specific Treasury proposals. Presumably the Treasury views its proposal as a package, with the increase in the corporate tax burden being economically palatable only if accompanied by the recommended structural reforms. Such nuances may be lost, however, when Congress next gets around to searching for additional revenues.

The Economics of the Corporate Income Tax

Revenue Yield and Effective Tax Rate

One area of public misunderstanding about the corporate income tax concerns its revenue yield and the percentage of corporate profits that is taxed. A common view is that the corporate tax burden is a relatively small share of corporate profits, implying that significant increases in the corporate tax burden could be achieved without significantly reducing incentives for corporate business activities, and in particular that the burden has been virtually eliminated by the Reagan tax cut program. In fact, the issue is quite complex: there is a good deal of uncertainty over the effective rate of corporate taxation, and while there has been a reduction in that burden in recent years, much of it has not been the result of the Reagan tax program.

The revenue yield of the corporate income tax has fluctuated sharply in the past few years, partly in response to cyclic fluctuations in corporate profits and partly owing to legislation. The peak year for receipts from the corporate income tax was fiscal year 1979, when the tax raised $66 billion. The recession and the 1981 tax reductions caused corporate receipts to fall to $49 billion in fiscal year 1982 and $37 billion in 1983; however, the economic recovery and the 1982 tax increases caused corporate receipts to rise back to $57 billion in fiscal year 1984. The Congressional Budget Office predicts a gradual rise in corporate tax revenues to $96 billion by fiscal year 1989. Thus, the revenues raised by the corporate income tax are still large enough to be of macroeconomic significance. Suggestions that we might as well repeal the tax because it raises little revenue are simply not in line with the facts.

In relation to the size of the economy, however, corporate tax revenues are very much smaller than they were in the 1950s and 1960s, particularly during the Korean War, when corporate income tax revenues were augmented by an excess profit tax. In fiscal year 1952, for example, corporate receipts peaked at 6.3 percent of gross national product, and during the sixties they hovered around 4 percent of GNP. These numbers may be compared to corporate tax shares of 1.1 percent of GNP in 1982, 1.6 percent in 1984, and an estimated 1.8 percent in 1989. This decline over the past several decades in the percentage of GNP raised in corporate tax revenues has been used to justify proposals to raise more revenue from the corporate sector, and some have argued that corporate tax increases can be the centerpiece of a major deficit reduction package. This sort of "restoration" of corporate tax receipts, along with defense budget cuts, was essentially the budget program of the Reverend Jesse Jackson in the 1984 presidential campaign.

Unfortunately, a restoration of the corporate income tax to its position in the 1960s, much less the 1950s, is unrealistic because corporate profits before tax are a much smaller share of GNP than they were in the 1950s or 1960s. One measure of corporate profits is contained in the national income accounts, where reported profits are adjusted for the effects of inflation and for accelerated depreciation. Corporate profits, after adjusting for the effects of inflation on inventory valuation and capital cost recovery (the so-called inventory valuation and capital consumption adjustments in the national income accounts), have declined from an average of 10.4 percent of GNP in the 1950s and 10.2 percent of GNP in the 1960s to 6.1 percent in the 1980s. Thus, in relation to the size of the economy, the potential corporate tax base is only about 60 percent as large as it was in the 1950s and 1960s. Restoring corporate tax receipts to 60 percent of their 1960s' share of GNP, or to about 2.4 percent of GNP, would involve a corporate tax increase in fiscal year 1989 of about 0.6 percent of GNP, or about $30 billion. This is approximately what is being recommended by the Treasury Department. It would be a significant contribution to deficit reduction, but it would still leave the United States with unacceptably large deficits.

When corporate taxes are measured as a percentage of the profits that corporations report to their shareholders, the resulting low effective tax rates also contribute to the perception that corporations are undertaxed. A study by the congressional Joint Committee on Taxation (1984) of the annual reports of several hundred large corporations pegs the average effective tax rate (current taxes paid to the United States divided by income from U.S. sources) at only 16.7 percent in calendar year 1983.

However, one problem with such studies is that corporate financial statements often overstate the corporation's true economic income. Financial statements tend to overstate real profits during periods of inflation primarily because depreciation costs are measured based on the original cost of the asset. As a result, they tend to understate the actual diminution in the value of the asset, measured in current prices, during years subsequent to its purchase. Moreover, first-in, first-out accounting for inventories tends to create inventory profits that are not accompanied by additional cash flow. A partial offset to these tendencies of financial accounting to overstate real corporate profits is the understatement that arises because reported corporate profits fail to include the reduction in the real value of indebtedness that results from inflation. Studies are not available to quantify the corporate effective tax rate based on inflation-adjusted profits, but it is likely to be significantly higher than that estimated in the Joint Committee study, particularly during years after there has been considerable inflation. (The Financial Accounting Standards Board has recently dropped the requirement that corporations publish the data with which such an analysis could be made in the future.)

A common misperception about the corporate income tax is that it has been gutted by the Reagan tax cut program. Actually, while the 1981 tax cuts did reduce corporate income tax receipts by about half, the 1982 and 1984 tax increases, which predominantly raised business taxes, offset almost all of the 1981 tax cut. For fiscal year 1987, for example, the 1981 tax cut for corporations amounted to an estimated $52.2 billion, but the 1982 tax increase amounted to $39.5 billion and the 1984 tax increase amounted to another $7.9 billion. Hence, the net corporate tax cut for the three bills combined is estimated to be only about $5 billion for 1987. The Treasury proposals, therefore, would mean a sizable tax increase for corporations not only relative to present law but also relative to pre-1981 law. Actually, a larger real effective corporate tax cut during the Reagan administration has resulted from the decline in inflation, which has reduced the overstatement of real taxable income that results from failure to make appropriate inflation adjustments than from the legislated tax changes.

Incidence

Who bears the burden of the corporate income tax? The corporate entities themselves cannot bear a tax burden; some individuals must bear it. Just who these individuals are is an important issue in assessing the desirability of any given level of corporate taxation.

One common view is that the corporate income tax is pushed forward to consumers as higher prices. This might occur if corporations determine the prices of the products they produce as some kind of mark-up over average costs, including tax costs. If consumers in fact bear the burden of the corporate income tax, there is a strong case for replacing the tax with direct, simpler taxes on consumers, which is why this particular theory of the incidence of the corporate income tax is frequently argued by business leaders and their representatives who would prefer reduction or elimination of the corporate income tax.

This theory that corporate income taxes get reflected in higher consumer prices is not very helpful for two reasons. First, it is very likely not true as a short- or a long-run matter. In the short run, corporations presumably set prices so as to maximize their profits; if so, there is no reason why a tax that is computed as a percentage of profits should alter the price at which a corporation's profits from producing a particular product are maximized. Therefore, the theory that corporations respond to increases in the corporate income tax by raising prices implies that they were forgoing opportunities to increase their prices prior to the tax increase, a questionable proposition on which to construct a theory of tax incidence. Only in the case of regulated utilities, where corporate income taxes are included as a cost of service for rate-making purposes, is the notion that corporate income taxes are pushed on to consumers as higher prices consistent with short-run profit maximization. Second, even if corporate income taxes do lead to higher consumer prices, one must still assess the effect of the tax on the distribution of real incomes in the economy in order to determine the incidence of the tax. Even if the corporate income tax affects the price level, its impact on real incomes may not be changed.

The more or less standard economic theory of the incidence of the corporate income tax, the Harberger model, is that in the short run it reduces after-tax profits, the burden of which falls on the corporation's shareholders and presumably is immediately reflected in lower stock prices (see Harberger 1962). However, investors will not permanently be willing to accept a lower rate of return on their investment in corporate stock than is received by investors in other activities. Therefore, the impact of the tax is to shift investment out of the corporate sector and into the noncorporate sector, causing the before-tax rate of return on corporate investments to rise and that on noncorporate investments to fall until the after-tax returns in the two sectors are once again equalized. In this case, the burden of the tax is shifted away from corporate shareholders to the broader class of owners of capital generally.

However, several qualifications must be made to this theory. First, to the extent that corporate profits represent something other than a competitive return to capital invested in the corporate sector, such as profits derived from market power, the tax will fall directly on the shareholders. Second, if the additional burden on capital reduces capital formation and, ultimately, productivity growth, some or conceivably all of the burden of the tax will eventually be shifted to workers in the form of lower real wages. Third, corporate and noncorporate investments may not be perfect substitutes in the portfolios of investors, in which case the rates of return in the two sectors need not be equalized and the shareholders will bear more of the burden than in the Harberger model. Finally, to the extent that corporations can respond to the tax by increasing their use of debt finance, which produces deductions for interest payments, there will not be as large a decline in corporate investment as is implied in the Harberger model and more of the burden of what tax remains after the response will be borne by the shareholders.

To sum up, then, there is considerable uncertainty over just who bears the corporate tax despite years of academic discussion of the subject. In the short run and intermediate run, some combination of the shareholders and owners of capital generally appear to bear the tax burden; over the long run, however, some of the burden may be borne by workers.

Distortions Caused by the Tax

The economic case against the corporate income tax is essentially that it creates economic inefficiency that is large in relation to the amount of revenue raised by the tax. Some of these distortions are inherent in the concept of the tax itself; others relate only to the particular version of the tax that the United States presently uses and could be reduced or eliminated through tax reform. The distortions relate both to what investments are made in the economy, to how those investments are financed, and to where investments are made. (Investment is used here to mean expenditures by business on productive assets, including purchases of plant and equipment, research and development, inventory accumulation, drilling of oil and gas wells, exploration for and development of mineral resources, and so forth.)

Allocation of Investment. In order to assess how the corporate income tax affects incentives to make particular kinds of investments, one should look not only at the way the tax applies to profits from investments but also to the way costs of making investments are recovered under the present tax rules. For example, when equipment is purchased, the taxpayer receives an investment tax credit and accelerated depreciation deductions but must pay tax on the income earned in connection with the equipment. When real estate is purchased, there is no investment credit, and depreciation is slower, but the gain on the sale of the building may be treated as a long-term capital gain in certain cases even if depreciation deductions had previously been deducted against ordinary income. The net effect of the tax system on incentives to make these investments will depend on the overall effect of all of these provisions, which in turn may depend on the taxpayer's particular tax situation.

There is not yet a consensus on just how the existing capital cost recovery and other tax rules distort choices between different kinds of investments, but it appears that there are significant distortions. These could be reduced by creating a more uniform system of capital cost recovery. Both the Treasury tax reform proposal and the Bradley-Gephardt tax reform bill attempt to do this.

To some extent, the distortions inherent in the existing capital cost recovery rules are intended by Congress. However, much of the distortion is entirely unlegislated. Some distortion arises because the law treats investments with different economic lives under the same tax rules. Some arises because the value of any given array of cost recovery provisions depends on the rate of inflation, so that a set of rules intended to be nondistortionary or to create a particular set of distortions can have unintended impacts if the inflation rate changes unpredictably.

A second, related set of distortions concerns just which corporations in the economy make particular investments. Under the present system, corporations face a wide array of marginal tax rates. (The marginal tax rate is the change in tax liability that results from an additional dollar of income or deductions, expressed as a percentage.) Some corporations pay a marginal rate of 46 percent on a more or less permanent basis. Smaller firms pay graduated marginal rates of 15, 18, 30, or 40 percent. Firms with net operating losses pay a zero marginal tax rate. Firms whose investment credits must be carried over because they exceed 85 percent of tax liability pay a marginal rate of 6.9 percent.[1] In other cases, timber companies can face marginal tax rates of 28 percent because income from the cutting of timber is treated as a long-term capital gain, and mining companies can face rates of 23 percent owing to the net income limitation on percentage depletion. Furthermore, the marginal tax rate applicable to a particular firm can change from year to year as its circumstances change. The net effect of all this is that different firms can face very different incentives to make the same investment as a result of their being in different tax positions, in which case the investments will not necessarily be made by those firms which can most productively use a given item of plant or equipment.

The most powerful unintended tax biases in the choice of investments occur when a firm can anticipate changes in its marginal tax rate during the life of an investment. For example, the tax system strongly encourages investments in situations in which a firm has a high marginal tax rate during the years it is receiving cost recovery deductions and a low rate in subsequent years when it is earning taxable income from the asset. Conversely, the tax law discourages investments in the opposite situation, when a firm has a low marginal tax rate when cost recovery deductions are received and a high rate later. Frequently, firms in this situation lease equipment from other firms who can make fuller use of the tax benefits.

Distortions in Financing. The tax system also creates major distortions in the way firms finance their investments. It discourages the use of the corporate form of business organization. It encourages debt over equity financing for corporations with high marginal tax rates because interest payments on debt are deductible while dividends paid to equity investors are not. The additional use of debt increases the riskiness of a corporation's financial structure, which can reduce its willingness to take risks in its operations. Also, the riskiness of the nation's financial structure is increased as each individual corporation increases its leverage, since each lender's riskiness increases as its borrowers become more highly leveraged. The growth of leveraged buy-outs in recent years represents the extreme form of this largely tax-induced disortion.

Two features of the tax system discourage use of the corporate form: the corporate income tax and the fact that corporations cannot pass their losses on to shareholders. For certain small corporations, these disadvantages are largely eliminated (although losses cannot be deducted in excess of basis) through use of Subchapter S.

However, for corporations with low marginal tax rates, issuing debt is unattractive because the interest deductions are not as useful. For these firms, issuing preferred stock is an attractive substitute for debt finance. The fixed dividend payments on the typical preferred create a pattern of payments similar to the interest payments on long-term debt, or adjustable rate preferreds can be used by corporations who would like their payments to approximate the interest payments on short-term debt. Issuing preferred stock has a tax advantage, relative to issuing debt, for a firm with a low marginal tax rate because a corporation that owns preferred stock can exclude 85 percent of dividends received and is, therefore, willing to accept a lower pre-tax rate of return on the preferred stock than on debt of the same issuer. For such a low-tax firm, selling preferred stock to corporations is more advantageous than issuing debt because the lower yield on the preferred more than compensates for the loss of interest deductions.

Deciding on the best form of financing becomes an even more complicated problem when a corporation is uncertain about what its marginal tax rate will be in the future. Consider, for example, a corporation which currently has a low marginal tax rate but may have a high rate in the future. Should it issue debt or preferred stock? One solution to this dilemma is a preferred stock that is exchangeable into straight debt at the option of the issuer, enabling the issuer to make the switch when his tax situation changes.

Another financing choice that may be influenced by the tax law is the decision on whether to distribute earnings to shareholders as a dividend. In the case of individual shareholders, dividends are taxed as ordinary income, while only 40 percent of long-term capital gains is taxed. Thus, for these shareholders dividends are a relatively unattractive way of receiving a return from an investment in corporate stock. Far more attractive are the long-term capital gains that may arise if the corporation is acquired at a price that reflects the value of retained earnings, the corporation is liquidated, or the stock price rises because the corporation is using its undistributed earnings to repurchase its own stock. Each of these techniques for distributing corporate earnings to shareholders in forms other than straight dividends has grown in recent years. The tax incentive for earnings retention (but not for stock repurchases or liquidations) distorts the allocation of capital in the economy toward those firms that already have earnings to retain.

However, when corporations own stock in other corporations, the incentives are different. The 85 percent dividends received deduction means that intercorporate dividends are taxed at a rate of 6.9 percent, compared to the corporate long-term capital gains rate of 28 percent.[2] Therefore, corporate shareholders would prefer to receive dividends. Only tax-exempt shareholders are indifferent to the tax consequences of dividends and capital gains.

The tax law also frequently encourages mergers and acquisitions. There are tax savings for taxable, noncorporate shareholders from acquisitions that cause cash to be distributed from corporations in a manner that is treated as a long-term capital gain rather than a dividend. Mergers can also be used to transfer net operating loss or credit carry-overs from a corporation that has little prospect of using them to a corporation that can use them.

Finally, the decision whether to buy or lease plant or equipment is strongly influenced by tax considerations. Firms which cannot make optimal use of tax benefits associated with the ownership of plant or equipment have a tax incentive to lease that property from someone who can use the tax benefits. Because the tax law requires that a certain amount of economic substance be built into leases, these essentially tax-motivated transactions can have effects on real economic outcomes. The attempt in 1981 to permit leasing without economic substance, so-called safe-harbor leasing, became a national scandal and was repealed in 1982.

These various distortions in financing that are caused by the tax system are often overlooked in discussions of tax policy, but that should not obscure their significance. The increase in the riskiness of financial structures may only be important in times of economic distress, but then it could prove to be very important. Similarly, the impact on economic growth arising from the misallocation of capital between firms can have an important cumulative impact over a period of decades, even if it is hard to detect any impact in any particular year.

Distortions in Source of Income. Yet another set of distortions arises from the different treatment of domestic income and income from foreign sources. Domestic investments are eligible for the investment tax credit and accelerated depreciation, while foreign investments generally are not, but this incentive for U.S. investment is counterbalanced by the fact that tax on income from foreign subsidiaries is deferred until the earnings are repatriated as dividends to the U.S. parent. This balance could shift unless a tax reform proposal is carefully drawn.

Also, the rules for computing the foreign tax credit can provide incentives for investment abroad in particular situations, such as in the case of a company with excess foreign tax credits that is contemplating an investment in a jurisdiction that imposes little or no local tax. Such an investment is effectively tax exempt because excess credits can be used to offset the U.S. tax on the profits. The Treasury tax reform proposal addresses this issue by no longer allowing excess credits from one country to be used against income from another country.

Proposals to Modify the Corporate Income Tax. The sum total of all of the distortions created by the corporate income tax, as it is currently applied in the United States, is probably quite significant in relation to the revenue raised by the tax. In many instances, recent proposals to reduce the burden of the tax, such as accelerated depreciation, have aggravated some of the distortions. Few are satisfied with the corporate income tax as it now stands, and there are numerous proposals on how to modify it.

Abolish the corporate tax? If the corporate income tax is so bad, why not simply abolish it? President Reagan casually suggested just this solution a couple of years ago, and the immediate adverse public reaction to his comment, which forced him to retract it the next day, gives one indication of the political reasons why the tax cannot simply be repealed. However, there are also some persuasive nonpolitical arguments for retaining some form of corporate income tax.

As long as there is a personal income tax, some mechanism is needed to prevent individuals from sheltering their income in the corporate form. This could take the form of taxing shareholders on their pro rata share of the corporation's retained earnings, although this presents significant technical problems which are discussed below. The present law contains two devices designed to limit use of the corporate form to avoid the individual income tax, the personal holding company tax and the accumulated earnings tax, but the proliferation of professional corporations suggests that these techniques are not getting the job done even with a corporate income tax as a backstop, much less without one. The existence of a corporate income tax with a tax rate not significantly different from the top individual tax rate is thus an important way of guarding the individual tax base from erosion.

A second reason to retain some type of corporate income tax is that abolishing it would provide unnecessary benefits to foreign investors. In many cases, foreign corporations investing in the United States may credit their U.S. corporate income tax against the income tax in their home country. Abolishing the U.S. tax would not change the overall tax burden of these companies but would simply shift collections from the U.S. government to some foreign government. While in the past direct foreign investment in the United States has not been of such macroeconomic significance that our tax policy need hinge on it, that situation may be changing.

Finally, abolishing the corporate tax would produce a giant windfall to shareholders of U.S. corporations, who invested under the assumption that their corporations' profits would be taxed. This is also an issue with any proposal to reduce corporate tax rates.

Cash-Flow Corporate Tax. One proposal to reduce the distortions arising from the corporate income tax, while avoiding the problems inherent in repealing it, is to convert the tax into a tax on corporate cash flow. This approach was recommended in 1978 by the prestigious Meade Committee in the United Kingdom. Under a cash-flow corporate income tax, all investments would be written off in the year they are made (they would be "expensed," in a word), business income would otherwise be determined under cash accounting, and borrowing would be treated on a cash-flow basis (borrowing would be added to the tax base and payments of interest and principal would be deducted). Note that this definition of cash flow differs from that used in certain kinds of security analysis, where cash flow sometimes measures corporate profits plus noncash expenses like depreciation and deferred taxes.

A cash-flow corporate tax, it is argued, would impose essentially no burden on new investment: the tax saving from deducting the cost of the investment would tend to equal, at least in present value, the tax collected on the income received as a result of the investment. Because this would be true of all investments, the system would not distort the choice of which investment to undertake. Yet, unlike repeal of the corporate income tax, a cash-flow tax would permit some protection of the individual income tax base and would not create a windfall for existing shareholders.

However, the proposal to convert the corporate income tax into a tax on corporate cash flow still presents a number of problems. First, it would involve losing much of the tax revenue raised under the present corporate income tax unless tax rates were increased dramatically. (The revenue loss arises primarily from expensing investments in depreciable assets.) Second, the neutrality property of the system—the proposition that expensing leads to a nondistortionary zero effective tax rate on all types of corporate investment—hinges on the assumption that the marginal tax rate stays constant throughout the life of the investment. If tax rates change, owing to legislation, to corporations' having net operating losses, or to other reasons, then distortions creep back into the system because the tax saving from expensing the investment no longer equals the tax on the income from the investment. Moreover, in an expensing system, firms are likely to generate tax losses whenever they are sharply increasing their level of investment. Third, a cash-flow corporate tax would make it possible to avoid the individual income tax on activities that involve making investments that could be expensed under the corporate income tax but not under the individual income tax. For these reasons, a cash-flow corporate income tax probably only makes sense if the individual income tax were itself converted to a tax on consumed income, which also would involve expensing of investment and cash-flow accounting for business income.

There have also been proposals to move part of the way toward a cash-flow corporate tax by permitting expensing of all investments, or at least investments in equipment. Until recently, expensing of much equipment was the policy in the United Kingdom; however, their recent budget proposes to do away with expensing in exchange for dramatically lower corporate tax rates. Also, expensing was part of the House Ways and Means Committee's alternative to President Reagan's tax reduction bill in 1981.

Serious problems would arise if an expensing system were adopted without the other adjustments needed to implement a cash-flow tax base. The double dip inherent in expensing debt-financed investments where a deduction for interest is allowed would lead to a dramatic erosion of the corporate tax base (and the individual tax base if expensing of debt-financed investment were allowed under the individual income tax). Also, the distortion between firms with different marginal tax rates would become even more significant.

Corporate Integration. Another proposal to reform the corporate tax structure is to integrate the individual and corporate income taxes. This can be done in several different ways, some of which appear more promising than others.

Under the present system, integration already exists in the case of so-called S corporations, which must have fewer than thirty-five shareholders and meet certain other requirements. S corporations are exempt from tax but instead flow through profits (but not all losses) to shareholders. Extending similar treatment to all corporations is known as "full integration."

If it were feasible, full integration would be consistent with the goal of comprehensive income taxation—taxing corporate-source income once at the shareholder's tax rate. It would eliminate biases caused by the corporate income tax in the choice of financing. However, without reform of the capital cost recovery system, it would not address the issue of tax-induced distortions in the choice of which investments to make.

Unfortunately, technical problems make it likely that full integration will not be possible for large companies. Audit adjustments to the corporation's income—which would occur years after the return is filed—would have to be flowed through to shareholders' tax returns, which would be extremely cumbersome in the case of large corporations with tens of thousands of shareholders. Problems would also arise when corporations owned stock in each other, since each would need to know the other's taxable income in order to compute its own taxable income. Shareholders would have to make complicated basis adjustments each year to reflect the corporate retained earnings on which they have already been taxed in order to prevent double taxation of retained earnings when a capital gains tax is paid upon sale of the stock.

A more limited approach to integration would be to deal only with the double taxation of dividends. Most other industrial countries eliminate some or all of such double taxation—about half in the United Kingdom and France and all in West Germany. Dividend relief, as this approach is called, can be done in either of two ways: allowing corporations to deduct the dividends they pay, or allowing shareholders to claim a tax credit for their pro rata share of the corporate income tax paid by the corporation whose shares they own. Substantively, these approaches are identical, but they present very different faces to lay observers.

Eliminating double taxation through either a dividends-paid deduction or a shareholder credit would not necessarily eliminate distortions in the choice of financing or the choice of business organization, but it would alter them significantly and probably improve them. The present bias in favor of earnings retention would be reversed: retention would be discouraged relative to dividend pay-out because there would still be a double tax on retained earnings—a corporate income tax and a capital gains tax arising from the higher stock prices that would result from retention. If corporations responded to the new regime by distributing all of their earnings, then the system would be neutral between debt and equity finance, since both the dividends (which would then be the only predictable return to equity investments) and interest would be deductible. Without double taxation, there would no longer be as large a bias against corporate-source income; the remaining bias would be that noncorporate entities could pass through losses to investors, while corporations could not.

If double taxation were eliminated, the tax incentive would be for corporations to distribute all of their taxable income as dividend payments. As long as the corporate tax rate is not significantly below the top individual rate (so there is little rate advantage to accumulating retained earnings in the corporate form) there is likely to be a dramatic reduction in retention of taxable earnings in a world without double taxation, in which case there would be little need even to consider addressing the technical problems of full integration. Very likely, in a world where tax considerations encouraged dividend payment, firms would make greater use of dividend reinvestment plans in order to entice shareholders voluntarily to provide the funds previously provided by earnings retention.

In choosing between the two approaches to dividend relief, several considerations are relevant, ones largely relating to perceptions, not substance. A dividends-paid deduction is a tax cut at the corporate level; a shareholder credit is a cut for the shareholders, who tend to have high incomes. Even if the two proposals have the same ultimate substantive impact, public perceptions of who is getting the tax cut may be different. Corporate managers may view the dividends paid deduction as preferable. Other people's perceptions of the equity of the income tax may be influenced by the shareholder credits, which would disproportionately reduce the taxes of upper income persons, even though in fact those credits simply refund tax collected at the corporate level. It may take decades for the public to adjust its view of the corporate income tax so that it is perceived like withholding on wages and salaries, refunds of which do not create perceptions of inequity. The potential significance of such perception problems is indicated by the case of Jacques Chaban-Delmas, the former prime minister of France who was forced to resign in 1971 after "revelations" that he paid no income tax in 1967 because of France's shareholder credit. This case suggests that a dividends-paid deduction may be a safer choice.

However, the dividends-paid deduction approach creates some perception problems of its own. First, if corporations do respond by distributing all of their taxable income as dividends, then the deduction would zero out corporate income tax collections. Second, while it may be straightforward to deny a shareholder credit to tax-exempt or foreign shareholders, for whom there is no double taxation, if this is desired, denying them the benefits of the dividends-paid deduction would require imposing a special tax on their dividend income to offset the corporation's tax saving from the deduction of dividends paid to them. Such a tax might prove objectionable, particularly if it violated tax treaties or if the organizations involved viewed themselves as being tax exempt on constitutional grounds. On balance, the dividends-paid deduction approach seems more promising in spite of these problems because it is simpler. Also, there is a strong case on economic grounds for giving foreign and tax-exempt shareholders the benefits of dividend relief; otherwise, some of the benefits in terms of reducing distortions in financing would be lost.

The Treasury Department came to essentially this conclusion. It proposed a deduction for one-half of dividends paid, offset only by a withholding tax on dividends paid to foreigners if such a tax did not violate tax treaties.

Historically, the problem with dividend relief has not really been substantive, but political. The corporate managers who are the principal constituency for corporate tax reduction typically prefer rate reductions or more targeted investment incentives like the investment tax credit or accelerated depreciation. The bias in the system toward earnings retention, which dividend relief would eliminate, makes life easier for the managements of those corporations with earnings to retain, so they have generally been lukewarm toward dividend relief. A world in which all funds had to be raised externally, and all investments subjected to the tests of the marketplace, would be problematical for corporate managers, who would have to devote much more of their energy to raising funds. Congress rarely gives large tax cuts to persons who are lukewarm, since many more grateful recipients are available.

Indexing. As noted previously, one of the principal objections to the present system of corporate taxation is that the incentives provided, and the extent of the real tax burden, vary capriciously with the inflation rate. A very significant reform would be to adjust the definition of taxable income to avoid this result. Treasury proposed doing this in its November 1984 tax reform proposal.

The indexing plan proposed by Treasury involves several interrelated adjustments. First, capital gains would be measured after adjusting the cost, or other basis, of each asset by the rate of inflation between the purchase and sale. Second, the basis of assets for computing depreciation would be adjusted for inflation. Third, inventory accounting under the first-in, first-out (FIFO) method would be adjusted for inflation so as to exclude from the tax base inventory profits that simply reflected inflation. Finally, borrowers and lenders would adjust the amount of the interest that they deduct (in the case of borrowers) or include in income (in the case of lenders) to exclude that part of the interest assumed to represent inflation. This percentage would be determined annually by the Treasury based on prevailing market interest rates and inflation rates.

If adopted, the Treasury indexing suggestions would lead to a much more accurate measure of a corporation's real income than exists today. The only qualification to this statement relates to the proposed inflation adjustments for debt, which are approximations that could turn out to be wrong. The real problems with indexing are likely to be its complexity and the difficulty in convincing borrowers of the merits of the cutback in their interest deduction. While the basis adjustment for depreciation will not add much complexity, and the adjustment for inventory may be no more complicated than the present last-in, first-out system, the adjustments for capital gains and debt would add significant complexity to the system. A partial indexing approach, doing only the simple adjustments, has the disadvantage of adding new distortions to the system. For example, indexing depreciation but not borrowing would encourage leveraged purchases of depreciable assets by high-bracket persons. Thus, the issue boils down to a trade-off between accuracy in measuring the tax base and complexity, the resolution of which probably depends on one's predictions of the inflation rate. In a world of little inflation, indexing the definition of income is not worth the added complexity; in a world with a lot of inflation, indexing is essential to achieve a meaningful definition of income.

Broad-Based, Low-Rate Income Tax. Another approach to corporate income tax reform is to broaden the tax base to approximate the definition of income used in financial statements and to lower tax rates. This is the approach used in the Bradley-Gephardt tax reform bill, which lowers the corporate rate to 30 percent. To the extent that the base-broadening involves reducing or eliminating the preferences in the law for various industries and types of investments, the effort would reduce distortions. The impact of remaining distortions would be mitigated by the lower rate. The lower rate would reduce, but not eliminate, distortions in the choice of financing. A broader based, lower rate system would also reduce the disparities in marginal tax rates across firms and over time, thereby reducing the distortions arising from this feature of the present system.

The real case for the broad-base, low-rate approach, which in effect substitutes lower rates for the structural reforms like dividend relief and indexing, is a pragmatic one. Administrative considerations and political realities will inevitably mean that the corporate tax base will be imperfect from an economic standpoint. The lower the rate, the less inequity and distortion from any given imperfection. If the complexity of indexing and the corporate ambivalence to integration rule out these proposals, the broad-base, low-rate approach, with rates comparable to those of the Bradley-Gephardt bill, would satisfy the political and substantive requirements of a corporate income tax while greatly reducing the distortions arising from the present system.

Relation between Corporate and Individual Tax Reform. The decision about what to do about the corporate income tax should be constrained by whatever approach is taken to reform of the individual income tax. While much of the economic benefit from tax reform, in terms of reduced distortion, comes from changes in the taxation of corporate-source income, the political force driving reform will have to be the desire to simplify the system for the individual taxpayer, possibly reduce his tax burden (as in the Treasury proposal) and make the system more equitable between individuals.

To the extent that simplification for individuals means repealing tax preferences used by middle-income people, political considerations will dictate that some comparable base-broadening must occur for upper income groups and corporations as well. If a preference is repealed for individuals, like percentage depletion, politics will also dictate that it be repealed for corporations. Similarly, if tax rates are lowered for individuals, it would be desirable to make a corresponding reduction in the corporate tax rate. Any system with a corporate rate significantly higher than the top individual rate would leave open possibilities to arbitrage the two systems in ways that are possible to a limited extent today but could be more widespread if the overall rate structure encouraged them. These considerations appear to rule out a high-rate, narrow-based approach to corporate tax reform, such as the cash-flow corporate tax or a system that would permit expensing of all investments, unless the dialogue on individual income tax reform moved away from comprehensive income taxation and toward taxation of consumed income.

Corporate Tax Increase Proposals. In addition to the various proposals for reform of the corporate income tax, a number of suggestions have been offered on how corporate tax revenues might be increased as part of an effort to reduce the deficit. The usual laundry list of corporate base-broadening suggestions is available to include in a straight tax-increase bill, but there are also some interesting more generic proposals that merit some comment.

One suggestion is a corporate income tax surcharge, expressed as a percentage of tax liability. If a revenue increase for individuals takes the form of an individual income tax surcharge, then it may be difficult politically for Congress to resist doing the same to corporations. Such a temptation should be resisted, because a corporate surcharge, which is effectively a rate increase, would simply serve to exacerbate all of the deficiencies of the present system.

An alternative suggestion is a corporate minimum tax, which has the distinction of being one of the few proposals to have been endorsed by both President Reagan (in his fiscal year 1983 budget) and presidential candidate Walter Mondale. Under the minimum tax, corporations would compute their economic income and pay a tax equal to 15 percent of that amount if this minimum tax exceeded their tax computed the regular way. A similar tax, with a 20 percent rate, currently applies to individuals.

Under the minimum tax, which could raise more than $5 billion per year at a 15 percent rate and much more revenue at higher rates, the entire additional burden would be borne by industries which currently pay low taxes. Furthermore, it would seem that a tax at such a low rate would not impose such an onerous burden that economic activity would be adversely affected.

Despite these attractions and its bipartisan sponsorship, the corporate minimum tax also ought to be bypassed in the search for new revenues. It would add complexity. Also, by establishing a new tier of marginal tax rates at 15 percent, the minimum tax would aggravate the problem of disparities in marginal tax rates among firms and over time. Because firms can conglomerate in so many different ways, the minimum tax would create competitive imbalances. For example, copper mining is a low-tax industry, in those years in which the industry is making any profits to tax. Therefore, an independent copper company like Phelps Dodge would have to pay a 15 percent tax on its income, while Kennecott, a subsidiary of Sohio, a high-tax company, would continue to pay no tax on its copper operations because, for minimum-tax purposes, its taxes would be averaged with the higher taxes of its parent company.

Pressure on Phelps Dodge to merge with a high-tax company could prove irresistible.

A more promising approach to generic corporate tax increases is the Stark-Chafee bill, which provides an across-the-board cutback in corporate (and individual) tax preferences. The cutback is 20 percent in Representative Stark's version and 10 percent in Senator Chafee's. This is a more comprehensive version of the approach included in the 1982 tax bill, which scaled back a selected list of corporate tax preferences by 15 percent (increased to 20 percent in the 1984 Act). The across-the-board preference disallowance tends to focus the tax increase on those firms which make disproportionate use of tax preferences. However, it would avoid the problems of the alternative minimum tax because the cutback would apply across the board to all corporations and this approach would not create a new tier of marginal tax rates.

CONCLUSION

One serious problem with the existing tax system is simply the frequency with which Congress sees fit to change it. The accelerating pace of change is making the system increasingly unworkable as the individuals who must implement the law—taxpayers, practitioners, revenue agents, and regulation-drafters at the Treasury—fall farther and farther behind in their understanding of a changing law. It is becoming increasingly difficult for businesses to plan for the short term, much less the long term, because of uncertainty about the tax law. Whatever change is adopted should include some sort of informal moratorium on further changes in the rules. (A formal moratorium is not possible because, absent a constitutional amendment, one Congress cannot bind another.) The problems with the present system seem too serious to permit a freezing of the status quo, but if Congress ever did enact a system that a consensus of informed people thought to be reasonable, it would make sense to give people some time to make that system work. This is one problem with the Treasury's desire to separate tax reform from deficit reduction, for such a separation guarantees a second major tax bill following on after the initial reform bill and a prolongation of the present uncertainty over future tax rules. It also suggests that tax reform should be preceded by a public debate that develops a well-grounded consensus for the sort of reform that is to be implemented.

NOTES

1. For such a firm, an additional dollar of income means an additional tax before credits of 46 cents; however, 85 percent of that tax may be offset by investment credit carryovers, leaving an additional tax after credits of 6.9 percent.
2. The dividends received deduction will offset the tax on all but 15 cents of a $1 dividend. A 46 percent tax on 15 cents amounts to 6.9 cents.

REFERENCES

Harberger, Arnold C. 1962. "The Incidence of the Corporation Income Tax," *Journal of Political Economy* 70 (June).

Joint Committee on Taxation, U.S. Congress. 1984. *Study of 1983 Effective Tax Rates of Selected Large U.S. Corporations.* November 28, JCS–40–84.

5 INCENTIVES AND WINDFALLS IN CORPORATE TAX REFORM

Alan J. Auerbach

Over the years, the corporate tax has evolved away from a tax on corporate income. This in itself is neither good nor bad, but it does affect the implications of various tax reform proposals. It is important to distinguish the long-run benefits of a tax system from its transitional effects. The latter may include economic costs outweighing the former gains. Costly windfalls may absorb much of the potential benefit of moving to a tax system that might have been preferable de novo.

The historical decline in corporate tax collections is largely the result of increases in investment incentives that have had a number of effects in addition to stimulating investment. These include an increased distortion in the allocation of capital within the corporate sector, a more acute problem of tax losses, and a large predicted discount in the market value of corporate stock attributable to the differential tax treatment of new and old capital.

Several options for corporate tax reform have been proposed in recent years. These include repeal, integration with the individual income tax, and, more recently, tax simplification schemes that would lower the corporate tax rate while broadening the base by reducing investment incentives. Included in the last category are the Bradley-

Gephardt Fair Tax and the Treasury's proposals of November 1984.

All three approaches would lessen certain of the distortions present in the current tax system, but generally at high cost in terms of the windfalls occurring to the corporate sector in the form of higher market values coming from forgiveness of anticipated taxes. Modifications to existing proposals can lessen such windfalls. Two approaches to accomplish this are discussed in this chapter's conclusion.

The title of this chapter refers to reform of the corporate tax, rather than the corporate *income tax.* This omission is not accidental. The tax base for what is frequently called the "corporate income tax" is not corporate income. One may argue if it ever was, but whatever pretenses might have been maintained about taxing corporate income surely were dropped in 1981, when "depreciation" passed out of the tax code and "capital cost recovery" appeared.

This evolution of the corporate tax away from an income base is, in itself, neither bad nor good. No one favors income taxation because income is inherently the right thing to tax, but rather because a broad-based income tax would satisfy a number of key economic policy objectives reasonably well. It is not the only tax that could do so, nor, as will become evident from my remarks if it is not already so, does attachment of the word "income" to a tax necessarily endow it with these desirable properties. This is all the more true when it is *corporate* income of which we speak, for most economists are uneasy with the notion that corporations have income that is in some manner distinct from that of their shareholders.

Hence, I will eschew taking a position about whether the corporate tax should be converted to a tax on corporate income, preferring to focus on specific effects of the current system and a variety of proposed alternatives, including the recent Treasury plan. I wish to emphasize, in particular, the difference between evaluating a tax system de novo and as an alternative to another tax system already in place. Costly (in terms of tax revenue) windfalls may absorb much of the potential benefit of moving to a tax system that might have been preferable in the first place. Old taxes may not be good taxes, but they enjoy a certain crucial advantage in this respect. The solution need not be to forsake tax reform, but perhaps to proceed somewhat more imaginatively.

SALIENT FEATURES OF THE CORPORATE TAX

To discuss the economic effects of the current corporate tax system that require attention, it is useful to summarize the features of the tax that lie behind these effects. We arrived at the current state with some purpose, and when considering reforms it is important to remember why at least some of us thought past "reforms" were the right ones to make at the time.

James Wetzler records in Chapter 4 of this book the decline in corporate tax revenues through the years. At their low point in fiscal year 1983, they amounted to little more than 6 percent of federal revenue, compared to 15 percent as recently as fiscal year 1978 and well over 20 percent in years past.

This decline has been accomplished primarily through the use of investment-related tax incentives, particularly the Investment Tax Credit and accelerated depreciation. The statutory corporate rate applicable to almost all of the aggregate corporate tax base, 46 percent, is only 6 percentage points below its 1950s level. Compared to the top bracket of the individual income tax, it has come down very little in the past twenty years.

Investment incentives have been the preferred vehicle for tax reductions for several reasons. First, in recent years, when inflation was a more serious problem than at the moment, the absence of price level indexing made depreciation allowances lose their value over time. Short of introducing such indexation, an increase in the value of depreciation deductions through acceleration was a more direct solution to the problem than a tax cut. Second, there lurks in the minds of many the view that investment in fixed capital is socially more desirable than investment in nondepreciable assets such as land and inventories. This view often distinguishes among types of fixed capital as well, favoring equipment over structures. I am unaware of any sound economic basis for this belief, but it is mentioned too often to be ignored as a contributing factor in the adoption of certain provisions.

Finally, and to me most importantly, investment incentives have been favored because a variety of studies over the years have suggested that they have more "bang for the buck": a greater increase in business activity, especially investment, per dollar of forgone tax revenue. This is because the incentives are "targeted"; generally only newly purchased capital goods qualify, and "antichurning" rules have been used to prevent "old" capital from reappearing elsewhere as "new." Hence, compared to a corporate tax cut, firms with old assets are treated more harshly. But this treatment is irrelevent to the question of how much new investment is undertaken; it is primarily a question (not at all an unimportant one) of income distribution or, if one prefers, fairness. Tax revenue reductions that go to reduce the tax burden on income from existing assets have essentially no impact on new investment, at least in theory. Empirical evidence seems to support this view, too.

If one views a corporation as generating income from new and old sources, the distinction between rate cuts and investment incentives is that the latter excludes old sources from benefiting. The efficiency argument is that these sources are there already; the investor does not need any incentive to build them. It would be as if the corporate tax were reduced only with respect to income derived from new investments. This identification would, of course, be impossible. Hence, such a policy of partial rate reduction has never been attempted.

The remaining feature of the corporate tax that lies behind both existing distortions and reform proposals is its independent existence alongside the individual income tax: the taxes are not integrated. Unlike ordinary partnerships, corporations do not pass through income or losses to their shareholders, but pay a separate tax which does not affect their shareholders' own tax liabilities. This results in the double taxation of corporate equity earnings, whereby dividends distributed out of after-tax corporate income are taxed once again at the shareholder level. It also introduces a nonneutrality between debt and equity finance, for with the deductibility of interest payments, corporate borrowers receive the same tax treatment as individual borrowers would.

ECONOMIC COSTS OF THE CURRENT SYSTEM

Several types of distortions can be traced to the current structure of the corporate tax in the United States. Any tax reform proposal should be evaluated according to how well it relieves the distortions in each category, and at what revenue cost.

Sensitivity to Inflation

One of the major arguments in favor of the Accelerated Cost Recovery System introduced in 1981 was that acceleration of depreciation allowances was needed to offset the reduced real value of depreciation allowances caused by inflation. By adopting the indirect remedy proposed by the Reagan administration, Congress loaded a gun that was set off by the government's successful anti-inflation policy. James Wetzler, quite correctly, states that the most important tax policy toward investment since 1981 has been this increase in the real value of depreciation allowances. According to my own calculations, new equity-financed corporate investments in equipment qualifying for the Investment Tax Credit now face a negative tax rate, regardless of the particular type of asset involved. This means that the combined value of Investment Tax Credits and depreciation allowances associated with a typical investment in equipment exceed, in present value, the anticipated corporate taxes the investing company will have to pay on the profits generated by the investment. The tax rates facing equipment are even more negative than they were in 1982, before the Tax Equity and Fiscal Responsibility Act introduced a 50 percent basis adjustment for the Investment Tax Credit.

Overall, I estimate that new equity-financed, fixed corporate investment faced an effective marginal tax rate of just over 19 percent in 1984. However, if the inflation rate rose permanently by only two percentage points, this number would have risen to nearly 25 percent. If inflation dropped two percentage points, the effective tax rate would fall to just over 12 percent. This represents an average increase in the effective tax rate of over three percentage points per every percentage point increase in the inflation rate. Given the range of inflation rates experienced even over the past four years, this represents a built-in potential for substantial fluctuations in the incentive for business to invest.

There are additional distortions to corporate financial decisions associated with an increase in the inflation rate, since full deductibility of nominal interest payments overstates corporate borrowing costs. At the individual level, interest income is likewise overstated. Since bondholders are likely to be in lower tax brackets than corporate borrowers, the net impact is to encourage corporate borrowing.

Interindustry Distortions

With a high statutory corporate tax rate, few capital recovery classes based only very loosely on productive asset lives, and a sensitivity of the tax burden to inflation, one should have little reason to anticipate that the effective tax rates facing different assets should be similar. They aren't. Table 5–1 gives my estimates of the effective tax rates

Table 5-1. Effective Tax Rates (%).

Asset Class		1984 (Current Law)
1.	Furniture and fixtures	−9.1
2.	Fabricated metal products	−7.9
3.	Engines and turbines	−7.1
4.	Tractors	−12.7
5.	Agricultural machinery	−8.2
6.	Construction machinery	−13.4
7.	Mining and oilfield machinery	−12.9
8.	Metalworking machinery	−9.9
9.	Special industrial machinery	−8.6
10.	General industrial machinery	−9.9
11.	Office and computing machinery	−21.0
12.	Service industry machinery	−12.9
13.	Electrical equipment	−9.6
14.	Trucks, buses, and trailers	−19.5
15.	Autos	−28.9
16.	Aircraft	−14.2
17.	Ships and boats	−6.8
18.	Railroad equipment	−6.3
19.	Instruments	−11.6
20.	Other equipment	−11.6
21.	Industrial buildings	39.7
22.	Commercial buildings	35.9
23.	Religious buildings	33.7
24.	Educational buildings	33.7
25.	Hospitals	35.4
26.	Other nonfarm buildings	42.5
27.	Railroads	22.6
28.	Telephone and telegraph	27.1
29.	Electric light and power	26.2
30.	Gas	12.2
31.	Other public utilities	14.4
32.	Farm structures	35.5
33.	Mining, shafts, and wells	22.2
34.	Other nonresidential structures	37.4
	Aggregate	19.4

facing new, equity financed investments in different types of corporate assets, and Table 5−2 gives industry-level effective tax rates based on the capital stock composition for representative firms in each industry. While some industries face tax rates on fixed capital investment close to the statutory rate of 46 percent, others face negative tax rates, in the aggregate. Again, this means that the typical firm in such an industry will generate tax deductions and credits in excess of future tax liabilities associated with the new fixed capital investments that they make, even if they finance none of the purchases by borrowing.

This variation in the effective tax rates facing different firms according to the assets they purchase distorts the allocation of capital in the corporate sector, and this distortion has the social cost of a reduction in national income. From a social viewpoint, we would like investors to choose those projects with the highest rate of return, before tax. But from the perspective of the rational investor, it is the after-tax return that matters. Hence, projects with low effective tax rates may be chosen even though they have relatively low before-tax returns, because a greater fraction of these returns may be kept by the investor. The greater the variation in effective tax rates, the more capital will be directed toward low-yield projects with favorable tax attributes. Were capital reallocated, more before-tax income could be produced with the same capital or, alternatively, the same before-tax income could be produced with less capital.

According to my own estimates, the percentage of the fixed corporate capital stock that is being "wasted" because of this misallocation now exceeds 3 percent. If this disortion were completely removed, it would be as if the corporate sector received a one-time grant of $75 billion for the purchase of additional capital goods.

A second major source of distortion comes from the misallocation of capital between the corporate sector, as a whole, and the noncorporate sector, primarily housing. The tax treatment of investments in owner-occupied housing is substantially more generous, even under current law, than the treatment of corporate investment. Not only is there no second level of taxation, but there is no first level, either. The income from owner-occupied housing consists of inputed rent, which is not taxed, and capital gains, which are effectively not taxed, when account is taken of the provisions allowing rollover and old-age exclusion of gains.

As a result, various studies have suggested that distortions, as big as those that exist within the corporate sector, are present because of this differential treatment of corporate and noncorporate investment. Such

Table 5–2. Effective Tax Rates, by Industry (%).

Industry Number		1984 (Current Law)
1.	Food and kindred products	23.8
2.	Tobacco manufacturers	20.3
3.	Textile mill products	19.1
4.	Apparel and other fabricated textile products	21.8
5.	Paper and allied products	12.4
6.	Printing, publishing, and allied industries	25.8
7.	Chemicals and allied products	14.9
8.	Petroleum and coal products	31.3
9.	Rubber and miscellaneous plastic products	13.8
10.	Leather and leather products	25.0
11.	Lumber and wood products, except furniture	20.9
12.	Furniture and fixtures	26.1
13.	Stone, clay, and glass products	20.2
14.	Primary metal industries	23.1
15.	Fabricated metal industries	18.7
16.	Machinery except electrical	20.1
17.	Electrical machinery, equipment and supplies	20.5
18.	Transportation equipment, except motor vehicls and ordinance	27.9
19.	Motor vehicles, and motor vehicle equipment	14.8
20.	Professional photographic equipment and watches	23.2
21.	Miscellaneous manufacturing industries	21.9
22.	Agricultural production	7.8
23.	Agricultural services, horticultural services, forestry and fisheries	3.5
24.	Metal mining	32.1
25.	Coal mining	8.8
26.	Crude petroleum and natural gas extraction	28.5

27.	Nonmetallic mining and quarrying, except fuel	1.0
28.	Construction	−5.6
29.	Railroads and railway express service	9.3
30.	Street railway, bus lines, and taxicab service	−20.2
31.	Trucking service, warehousing, and storage	−13.8
32.	Water transportation	−6.4
33.	Air transportation	−12.7
34.	Pipelines, except natural gas	13.4
35.	Services incidental to transportation	−0.1
36.	Telephone, telegraph, and miscellaneous communication services	12.9
37.	Radio broadcasting and television	23.3
38.	Electric utilities	20.7
39.	Gas utilities	11.6
40.	Water supply, sanitary services, and other utilities	37.1
41.	Wholesale trade	8.3
42.	Retail trade	23.5
43.	Finance, insurance, and real estate	34.9
44.	Services	18.4

distortions have undoubtedly been lessened in recent years as the effective tax rates facing the corporate investment declined.

Tax Losses

The U.S. tax system does not allow firms with tax losses to get a refund. They must either carry losses back against previous taxable income, or forward with the hope of generating taxable income sufficient to absorb the losses from their tax deductions. Why the tax system is structured this way, like the lack of price level indexing, may be traced to conditions that once held but no longer do. In this case, the condition was the use of a tax base that roughly approximated economic income. In such a situation, one could at least make a reasonable argument that firms with losses were, in general, less successful firms. Now, however, the situation is quite different. With Investment Tax Credits and accelerated depreciation, a successful firm can generate tax losses simply by growing quickly.

To the extent that the lack of refundability for losses was intended to limit the use of tax benefits to successful, ongoing enterprises, the generation of losses by growth is clearly an accidental result. It leads to a situation in which firms with different tax status face different incentives to invest, with a bias in favor of large, integrated organizations that can find sufficient sources of taxable income to make current investment incentives usable.

The solution to this problem introduced in 1981 was "safe-harbor" leasing, which facilitated the transfer of tax benefits associated with new investments. But, at least in part because the transfer of such tax benefits are not limited to otherwise profitable companies, safe-harbor leasing encountered criticism on both political and economic (not to mention legal) grounds. There is at present no alternative mechanism in place or contemplated by pending proposals.

Financial Distortions

As the corporate tax permits the deductions of interest, one would expect firms to finance a large fraction of their investments by borrowing. Yet firms in the United States do not borrow nearly as much as their counterparts in many other countries, notably Japan. There has yet to appear in the economics or finance literature a single theory capable of explaining this financial pattern. As a result, it is difficult to understand exactly how the current tax system's treatment of debt and equity affects behavior. As an example of the uncertainty, consider the impact on investment of the double taxation of corporate dividends. Common arguments suggest that the lack of deductibility of corporate dividends places an additional burden on firms to the extent that they use equity finance. But suppose the source of the equity funds in question is retained earnings, as is the case for most mature corporations. Then the double taxation argument breaks down, since there is an additional, favorable tax effect that occurs when the earnings are retained: dividend taxes at the shareholder level are reduced. Indeed, this reduction normally just offsets future paid-dividend taxes paid on the income from the new investment, assuming that the firm earns a normal rate of return on the investment. Put another way, the level of the dividend tax should not enter directly into whether the project is worth undertaking. This is not to say that there is no additional tax burden imposed by dividend taxation; just as in the case of corporate rate changes altering the tax burden on income from existing assets, taxes paid on current income are not necessarily relevant to the question of incentives for new investment. Lowering the tax burden on corporate dividends would undoubtedly lower tax collections, but its incentive effects are less clear.

OPTIONS FOR TAX REFORM

In the best of all possible worlds, we should seek changes that would simplify the tax system, reduce distortions within the corporate sector and between corporate and noncorporate sectors, reduce the sensitivity of the tax system to inflation, and enhance the overall incentive to invest. Most important, all this should be achieved at a minimum tax revenue cost, since there is little disagreement that major tax cuts would constitute inappropriate fiscal policy in the near future.

Because James Wetzler has already explored many aspects of these tax reform proposals (in Chapter 4), I will focus primarily on the revenue question: What overall incentive to invest does each reform provide, given the revenue cost involved? Of course, the other respects in which proposals differ are sometimes closely related to the revenue effects, so I will also explore ways in which programs that would be desirable from other perspectives (for example, reduced distortions or simplicity of administration) can be improved in terms of their revenue cost.

Corporate Tax Repeal

Despite the recent small proportion of tax revenue that the corporate tax has accounted for, repeal would still represent a substantial revenue loss to the Treasury. At the same time, it would, of course, eliminate many of the distortions present in the current system. With no taxation at all, the problems associated with losses, differential taxation, and inflation would disappear, and the distortion caused by the misallocation of capital between corporate and noncorporate sectors would be lessened by the reduction in the effective corporate tax rate. However, repeal of the corporate tax is simply an extreme version of a reduction in the statutory tax rate, which is an inefficient type of investment incentive because of the windfalls to owners of existing assets.

The magnitude of this windfall is quite large. The corporate tax system's investment incentive provisions have led to a situation where, in 1984, the existing stock of corporate capital should have had a stock market value of approximately two thirds of its replacement cost. This discount is solely related to the tax system's differential treatment of new and old capital. For example, a new asset in the five-year ACRS category receives investment credits and depreciation allowances worth, in present value, over 46 cents in after-tax dollar of investment. A comparable piece of equipment five years older, whatever its replacement cost, has no such deductions and credits. Therefore, the value of its future after-tax returns is lower by this factor, and the market price of the corporation that owns this type of asset should reflect the difference. For this example, the difference between the value of new and old assets would be over 46 cents. In the aggregate, it is somewhat smaller. One may think of this discount as representing the extra taxes that owners of ACRS property expect to pay because their assets are not new and have already received some or all of their investment incentives.

What would happen to this discount with the abolition of the corporate tax? It would disappear as well, representing a predicted increase in the value of corporate equity of approximately three-quarters of a *trillion* dollars. This is the present value of tax revenue being forgone through windfalls to existing capital. It is not only large in absolute magnitude, but large also *relative* to the total present value of corporate tax revenue that would be lost through repeal. The immediate implication is that corporate investment incentives could be improved substantially, or revenue losses reduced substantially, if such windfalls were reduced.

Given this situation, corporate tax repeal is probably best thought of as a benchmark against which to compare other, less extreme proposals.

Corporate Tax Integration

The view that corporations should be taxed as independent entities finds little support among economists. The full integration of individual and corporate taxes would seem the obvious remedy, with the corporate tax being converted essentially to a withholding mechanism for the individual income tax. Wetzler (Chapter 4) has outlined the technical difficulties that have been found when full integration was contemplated, and also the two "partial integration," or "dividend relief" schemes that have been proposed as partial reforms in the direction of full integration.

The Treasury, in its recent proposal, has opted for the dividends-paid deduction, but would allow a deduction equal to only half of dividends paid. This is the simpler of the two schemes but has its own problems. Under current law, the most serious would be the prevalence of companies without sufficient taxable income to make use of their deductions. However, Treasury would also increase the rest of the corporate tax base, thereby lessening this potentially serious problem.

The dividends-paid deduction is seen as a way of reducing the double taxation of dividends. Under full deductibility of dividends, earnings would be taxed only once, regardless of disposition, at the corporate rate if retained and at the individual rate if distributed. Under Treasury's plan, there would remain some taxation of distributed earnings at both levels.

What is least clear about the dividend relief approach, both Treasury's in particular and others in general, is what effect it will have on the incentives for corporations to invest. To the extent that new investment comes from retentions and dividends are taxed at a constant rate over time, I can see no improvement in the incentive for corporations to invest. Though future earnings will be taxed heavily, the funds currently used for investment become more expensive. It is essentially a wash, except for one major effect: taxes paid on dividends will go down. This makes "dividend relief" potentially even less efficient than corporate tax abolition as a way of providing improved investment incentives. I do not dispute that a reduction in taxes on dividends will occur, merely that it will have an appreciable impact on investment by companies obtaining their equity funds through earnings retention.

There is one aspect of the Treasury's plan that offers some promise in this regard. Instead of a straight switch to the 50 percent dividend deduction, the plan would offer a phase-in over five years, beginning with 30 percent deductibility in 1986 and increasing the percentage deductibility by 5 percent per year until 1990. This approach would have two positive effects relative to the straight implementation of the 50 percent dividend deduction. First, it would reduce the windfalls to existing capital, and associated tax revenue loss, by delaying the plan's benefits. Second, by having a schedule of declining effective tax rates on dividends over time, the system will offer an improved incentive to invest out of retentions *during the transition* period. This is because the cost of retained funds, after tax, will be going *up* while the effective tax on dividends is going down. The best way to take advantage of this is to retain funds when dividend taxes are high and distribute dividends when taxes are low. In such a case, the effective rate of tax on earnings distributed as dividends would actually be *less* than the effective corporate tax rate alone: the tax gain associated with the initial funds retained would more than offset the additional taxes and subsequent dividends coming from the investment.

For corporations whose equity comes primarily from the sale of new shares, more standard analysis of the effects of integration is appropriate. Such firms *do* face double taxation on their distributed earnings, because funds for investment come to the firm without any immediate tax benefit to shareholders. For such firms, dividend relief, such as proposed by the Treasury plan, would, indeed, lower the overall effective tax rate on equity earnings.

Tax Simplification Schemes

The Treasury's tax simplification plan, along with the Bradley-Gephardt Fair Tax plan would remove the Investment Tax Credit, decelerate depreciation allowances to approximate economic depreciation, and scale back other provisions beneficial to specific industries, such as oil depletion allowances. The third major proposal, the Kemp-Kasten "Fair and Simple Tax Act," or FAST, would be less radical in its changes from the current system. All three plans would lower the main statutory corporate rate: the Treasury to 33 percent, Bradley-Gephardt and Kemp-Kasten to 30 percent.

The plans differ in other respects, too, including the extent of price-level indexation, but in terms of the efficiency of investment incentives, perhaps the most important characteristics of the plans are reductions in the corporate rate combined with the reduction in the value of investment incentives.

Remember: A corporate rate cut is less efficient in its targeting of revenue losses toward new investment than either the Investment Tax Credit or the acceleration of depreciation allowances. This is because all sources of corporate income, not just new ones, benefit from the rate cut. It therefore follows that a revenue-neutral combination, like the Bradley-Gephardt plan, that lowers the corporate rate while doing away with investment incentives, is bound to increase the effective tax rate on investment income from depreciable capital. My calculations suggest that it does, and the similarity of the plans suggests that the same effect would occur under the Treasury's proposal.

This result is unfortunate, for there are many reasons to favor a program that reduces corporate rates and utilizes economic depreciation allowances. It lessens the effective tax rate on investments in nondepreciable assets, such as inventories, and lessens the dispersion of effective tax rates among depreciable assets. According to my estimates, the cost due to misallocation of capital *within* the corporate sector would essentially disappear under Bradley-Gephardt.

Yet, with a requirement (implicit or explicit) that a certain level of revenue be raised from the corporate sector, higher windfalls mean higher marginal tax rates on new corporate investment. My estimates suggest that about half the tax discount currently present in the value of existing assets would disappear under Bradley-Gephardt (with a comparable windfall anticipated under a similar proposal such as the Treasury's).

As with proposals that would integrate corporate and individual taxes, a simplified, broad-based corporate income tax with low rates would unquestionably be preferable to the current system, if one were setting out to design a tax system de novo and windfalls were of no concern. But this is not a fair description of the present situation.

WHICH WAY OUT? SOME CONCLUDING THOUGHTS

To criticize specific proposals is not to deny the existence of the problem that gave rise to them. The corporate tax needs reform. Many of the approaches discussed here would lessen certain distortions present in the current system, but windfalls seem to be endemic. The crucial question, in the current state of increased budget realism, is whether this unfortunate connection is inevitable. I would argue that it is not. Moreover, there is more than one direction in which the existing tax reform proposals could be adapted toward the end of reducing windfalls.

First-Year Capital Recovery

One possible solution is to make all of the changes in the corporate tax by restructuring investment incentives, without any change in the statutory corporate tax rate. Within this framework, one could then decide how high or low the effective marginal corporate tax rate should be, and adjust the depreciation schedule accordingly.

One variant of this approach would be cash flow taxation, discussed in Chapter 4. This scheme would replace the Investment Tax Credit and depreciation deductions under present law with the ability to expense all investments immediately. There would be a zero effective tax rate on all new investments, and depreciation deductions, since taken all in the first year, would not be sensitive to inflation.

A second variant that would produce higher effective tax rates but more revenue is a scheme I proposed with Dale Jorgenson (1980) a few years ago to grant discounted rather than full expensing, but again entirely in the year of an asset's purchase. Under this approach, an asset purchased for $1,000 might receive $700 or $500 of depreciation deductions in the year of purchase, with no subsequent deductions. As with a cash-flow tax, the first-year capital recovery plan would be insulated from inflation, and could be made neutral across asset classes.

The major problem in enacting plans of this type would be the problem of tax losses mentioned before. As with the current system, plans that offer accelerated deductions can present healthy firms with tax losses, making the deductions less valuable. Hence some solution, such as permitting businesses to carry forward their unused tax loss deductions with interest, would be desirable.

Phase-ins

It would also be possible to go directly to a broad-based, low-rate corporate tax, with reduced windfalls along the way, if the new system were phased in. Under the Bradley-Gephardt proposal, a gradual rather than immediate reduction in the corporate rate from 46 percent would, according to my estimates, actually stimulate more investment during the phase-in than if the rate were dropped immediately. This is because the biggest depreciation allowances would still, even under Bradley-Gephardt, occur in the earliest years, and firms would benefit by making investments in years with higher tax rates prevailing but lower rates anticipated.

At the same time, such a phase-in plan would increase tax revenue, by reducing the windfalls delivered to existing income. Even more investment stimulus would be provided if the switch from ACRS to economic depreciation deductions were also phased in, for there would be the added incentive to invest under the older, more generous scheme.

Conclusions

In this last section, I have tried to support the position that revenue-conscious, incentive-preserving reform of the corporate tax is possible and, indeed no more radical or complicated than the proposals currently in fashion. What constitutes a windfall, of course, depends on what is anticipated by actors in the market. This is but one of the many complications not considered in my discussion. But a consideration of transition issues is of great importance if tax reform is to be feasible in the current economic environment.

REFERENCES

Auerbach, Alan J. and Dale W. Jorgenson. 1980. "Inflation-Proof Depreciation of Assets." *Harvard Business Review* 58, no. 5 (September/October): 113–118.

6 TAX POLICY INFLUENCE ON VENTURE CAPITAL

Burton J. McMurtry

The purpose of this chapter is to examine the most significant economic policies affecting the venture capital process. Venture capital investment exhibits some fundamental characteristics. Venture capitalists:

- Finance new and rapidly growing companies
- Focus on businesses which enhance productivity, improve quality of life
- Purchase equity securities
- Add value by active participation with company management
- Take high risks with expectation of high rewards
- Have a long-term orientation—five to ten years

Although professional venture capitalists finance only a small fraction of the businesses started during any given year, they do provide financing for a disproportionate share of rapidly growing companies.

Many high-technology businesses are venture financed. A recent study at the Massachusetts Institute of Technology (Birch and MacCracken 1984) concluded that

> American high technology firms provide the basis for innovation and the

productivity tools for small companies that produce most of the new jobs, and for innovative larger companies to remain competitive in world markets. They do not themselves provide a major source of employment.

Job creation is an important part of the venture process as we will see later, but the net innovation and productivity increase may be as important or an even more important result.

Venture capitalists generally purchase equity securities; that is, they provide permanent capital. Venture capitalists usually have no opportunity to sell any of their securities in less than five to seven years, and they often hold securities for ten years or more. This is truly long-term investing.

Venture capitalists most often serve on the board of directors of companies in which they invest, and they work extremely closely with the management team. They try not to run companies, but most are very active participants.

Risks are high in the venture business, but so are the rewards. Later we will examine just how risky it is.

ECONOMIC IMPACT OF VENTURE CAPITAL-BACKED COMPANIES

A 1982 General Accounting Office study of seventy-two companies started by venture capital with investment of only $209 million during the 1970s found that by the end of the decade these companies had

- Combined annual sales totaling $6 billion
- An average annual growth of 33 percent per year for the last five years
- Created an estimated 130,000 jobs
- Produced $100 million in employee tax revenues
- Produced $900 million in export sales

In addition, these companies paid substantial corporate income taxes and conducted a disproportionately high level of research and development.

A recent study (Venture Economics 1985) of 405 companies founded from 1968 to 1980 provided the growth record shown in Table 6–1. Data on employee taxes and export sales from these companies are not yet available. The performance of these companies is dramatic, and it

Table 6–1. Economic Impact of Venture Capital-Backed Companies.

				Amounts in 1983 ($ Millions)					
		Number of Employees		Revenues		Corporate Taxes		R&D Expenditures	
Years since Founding	Sample Size	Aggreg.	Avg.	Aggreg.	Avg.	Aggreg.	Avg.	Aggreg.	Avg.
3	209	37,238	178	$2,696.6	$ 12.9	$ 81.3	$0.4	$424.9	$2.0
6	175	74,062	423	5,389.4	30.8	244.9	1.4	328.6	1.9
9	154	105,350	684	6,865.9	44.6	336.9	2.2	453.1	2.9
12	128	150,929	1179	10,682.1	83.5	578.1	4.5	687.8	5.4
15	71	128,391	1808	10,727.6	151.1	324.8	4.6	693.4	9.8

Source: Venture Economics, Wellesley Hills, Mass., preliminary data, 1985. *Note:* Data from a sample of 405 companies founded 1968 to 1980.

helps indicate why policies encouraging such investment should be aggressively pursued.

Just how much investment was required to produce these results? Investment data are available for a 92-company subset of these 405 companies as shown in Table 6–2. Note that the average company in year 3 has relied primarily on equity capital and is still in a position of cumulative loss (with negative retained earnings). Very little long-term debt is available to young companies, and that which is available is most often lease financing for standard capital equipment. Venture capitalists provided essentially all of the equity capital through year 3. Only a small amount was provided by the public market, since most companies cannot attract public financing at such a young age.

It is obvious from experience and from Table 6–2 that growth companies require continuing financing. Note that the average company required almost twice as much equity financing between its third and sixth year as it required during its first three years. Much of this capital comes from the public equity markets. Hence it is extremely important that that market be healthy.

As companies become profitable they are able to finance internally (through retained earnings) some of their growth, and they are able to attract long-term debt in addition to equity.

How much investment is required to produce a job or a particular level of revenue? Table 6–3 combines data from Tables 6–1 and 6–2. The cumulative external investment (equity capital plus long-

Table 6–2. Sources of Capital for Venture Capital-Backed Companies, Average Amount per Company in 1983 Dollars (Millions).

Aggregate Amount through:	Equity Capital	Retained Earnings	Long-Term Debt	"External" Capital[a]	"Total" Capital[b]
Year 3	$ 7.1	$ (0.3)	$ 1.6	$ 8.7	$ 8.4
Year 6	19.6	5.3	14.5	34.1	39.4
Year 9	20.9	11.6	15.9	36.8	48.4
Year 12	24.7	27.9	13.7	38.4	66.3
Year 15	41.6	22.1	16.9	58.5	80.6

Note: Data from a sample of ninety-two companies founded 1968 to 1980.
a. Equity capital plus long-term debt.
b. Equity capital plus long-term debt plus retained earnings.
Source: Venture Economics, Wellesley Hills, Mass., preliminary data, 1985.

Table 6–3. Productivity of Capital Investment (External) Average Amount per Company in 1983 Dollars (Millions).

Years since Founding	Cumulative External Investment[a]	Jobs	Investment per Job ($ 000)	Revenue	Revenue ÷ Investment
3	8.7	178	48.9	12.9	1.5
6	34.1	423	80.6	30.8	0.9
9	36.8	684	53.8	44.6	1.2
12	38.4	1179	32.6	83.5	2.2
15	58.5	1808	32.4	151.1	2.6

a. Equity capital plus long-term debt ("external" capital).

term debt) required to produce a job ranges from roughly $30,000 to $80,000, averaging about $50,000. This is efficient job creation by almost any standard and such investment should obviously be encouraged. Note also that $1.00 of cumulative external investment is required to produce between about $1.00 and $2.50 of annual corporate revenue. The significant decrease in the cost of job creation and increase in revenue production in later years is a result of the ability of the companies to finance a sizable amount of their growth through their profits retained in the business.

Profits almost always must be retained to finance growth, and as a result almost no rapidly growing companies pay dividends. Most venture capitalists wouldn't recognize a dividend if they received one.

The obvious conclusion is that growth companies need lots of capital, not just at start-up but all along. Strong equity and debt markets are needed. It is most critical that equity markets be healthy. If equity is not available, as was the case during the early and mid-1970s, then companies will not be created or be permitted to grow.

A VENTURE CAPITAL RISK/REWARD PROFILE

How risky is the venture capital business? Venture capitalists describe the business as risky, but many have very selective memories when it comes to quoting actual statistics. That is not surprising since if one clearly remembered all his losses it might be very difficult to make another positive investment decision!

A very informative unpublished study by Horsley Keogh and Associates of Pittsford, New York, examines the performance through 1983 of ten venture capital funds in which they had invested between 1972 and 1981. These funds made 525 separate investments decisions in 358 different companies. Table 6–4 shows the resulting performance. Note that roughly 26 percent of the investments lost money and consumed 34 percent of the capital. Another 25 percent of the investments produced only a return of the original capital; these investments actually "lost money" also since capital was tied up for years with no return other than the original capital. Roughly 50 percent then were effectively losses! Almost another 40 percent returned less than five times the original investment, whereas 5 percent returned five to ten times and another 5 percent produced a return of more than ten times.

It sounds as if these venture capital funds must have been failures, but in fact all were successful. The time-weighted rate of return of the composite of these ten venture capital partnerships was 35 percent for the ten-year period ended December 31, 1983. The rates of return varied between funds from a low of 16 percent to a high of 83 percent.

Venture capitalists experience substantial losses, but their successes are dramatic and can overpower the losses, as seen in these examples. Entrepreneurs and venture capitalists are driven by the opportunity to produce a great success and to be able to benefit financially from that

Table 6–4. A Summary of the Performance of Ten Venture Capital Funds ($ Millions).

Result	Number of Investments	Percentage of Investment	Cost	Cost (%)	Value	Value (%)
Below cost	135	25.7	99	33.8	26	3.2
At cost	131	25.0	40	13.7	40	4.9
1–5 × cost	203	38.7	122	41.6	307	37.3
5–10 × cost	29	5.5	19	6.5	129	15.7
Over 10 × cost	27	5.1	13	4.4	321	39.0
Total	525	100	293	100	823	100.1

Note: Based on ten successful venture capital partnerships founded from 1972 to 1981. Sample includes 525 investment decisions in 358 separate companies. Based on results through December 31, 1983.
Source: Horsley Keogh and Associates, Pittsford, N.Y. 1984

success. As can be seen from these data, great successes are infrequent.

RECENT HISTORICAL FLOW OF VENTURE CAPITAL

Venture Economics in Wellesley Hills, Massachusetts, has kept an excellent data base on the venture capital business since 1969. Table 6–5 shows the flow of funds into and out of the venture capital industry from 1969 through 1984 and includes information about the availability of public financing for young companies during that same period. The spaces between 1969–70 and 1977–78 help focus on the dramatic changes at those times in both the venture market and the initial public offering market. What could have caused those changes? Table 6–5 reveals just how dramatic the changes were.

As a practicing venture capitalist since 1969, it is clear to me that the most significant change-producing factors were changes in the taxation of capital gains. The maximum capital gains tax rate was 25 percent prior to 1969, but beginning in 1969 it was increased. By 1976 it stood at 49.1 percent. In 1978 the maximum rate was dramatically reduced to 28 percent and then to 20 percent in 1981. It is clear from my personal

Table 6–5. Venture Capital Industry Estimated Fundings and Disbursements ($ Millions).

Year	Net New Private Capital Committed to Venture Capital Firms[a]		Size of Total Pool	Estimated Disbursements to Portfolio Companies		Public Underwritings of Companies with a Net Worth of $5 Million or Less: Number	Amount
1984	$>4,000	↑	$>16,000	$>3,000	↑	(224)	$ 1,186
1983	4,500		12,100	2,800		(477)	3,671
1982	1,800		7,600	1,800		(113)	619
1981	1,300	$13,200	5,800	1,400	$11,650	(306)	1,760
1980	700		4,500	1,100		(135)	822
1979	300		3,800	1,000		(46)	183
1978	600	↓	3,500	550	↓	(21)	129
			Capital gains tax decrease				
1977	39	↑	2,500–3,000	400	↑	(22)	75
1976	50			300		(29)	145
1975	10			250		(4)	16
1974	57			350		(9)	16
1973	56	$ 466		450	$ 2,935	(69)	160
1972	62			425		(409)	896
1971	95			410		(248)	551
1970	97	↓	2,500–3,000	350	↓	(198)	375
			Capital gains tax increase				
1969	171		2,500–3,000	450		(698)	1,367

a. Total new private capital less capital withdrawals.
Source: Venture Economics, Wellesley Hills, Mass.

experience that these changes had an enormous impact on the attitudes of both entrepreneurs and venture capital investors.

It would be overly simplistic and probably wrong to take the position that the only factor driving venture capital availability is the tax rate on capital gains and the differential between tax rates on capital gains and ordinary income. It would be even more foolish to ignore such strong phenomenological evidence and argue, as some do, that capital gains tax rates could be raised and the differential eliminated without significant disruption to the number of startup companies and to the flow of equity capital to growth companies.

I have vivid memories of the venture capital climate in the mid-1970s. Four of us spent six months full time in 1973 and 1974 trying to raise $15 to $20 million for a new venture fund. If a prospective source told us there was even a 10 percent chance that they might invest, we got excited and went to see him. It was hard to find sources that would give us even a 10 percent chance. After enormous effort we were able to raise $19 million from seven institutions in 1974. This $19 million was one-third of the total flow into venture capital firms in 1974! We were glad we hadn't waited for 1975, which was the low-water mark at $10 million.

Before examining the sources of venture capital it is important to emphasize that the total amount of professionally managed venture capital is an extremely small factor in our overall economic picture. It represents only one-tenth of 1 percent of GNP and the approximately $3 billion invested in 1984 represents only about 50 percent of the annual research and development budget of the IBM Corporation. The venture capital industry is, nevertheless, a highly leveraged industry, for example, in job creation and in productivity. It is the envy of the free world and the subject of much investigation and imitation worldwide. Tax policies which might impede that process should be discouraged.

SOURCES OF VENTURE CAPITAL

Table 6—6 shows the sources of venture capital for independent private firms from 1981 through 1984. This represents 67 to 80 percent of the total capital committed for these years as shown earlier in Table 6—5. It is clear that pension funds and foreign sources are playing an increasingly important role.

At least two questions are obvious from Table 6—6. First, what attracts these investors, particularly the pension funds? Second, since such a large percentage of recently committed venture capital comes from either nontaxed or preferentially taxed institutions, why should venture capital be at all sensitive to tax rates? Exhaustive answers to these questions would require a much longer essay, but some observations can be made.

Investors, including pension funds, have been attracted by the usually high rates of return exhibited since 1978, for example, by the 35 percent per year average performance of the ten funds analyzed earlier. These high returns were made possible by the ebullient public equity markets that developed after the 1978 and 1981 capital gains tax

Table 6–6. Sources of Venture Capital Commitments (Independent Private Firms Only).

	Total Capital Committed ($ Millions)				Percentage of Total Capital Committed			
	1981	1982	1983	1984	1981	1982	1983	1984
Pension funds	$ 200	$ 474	$1,070	$1,085	23%	33%	31%	34%
Individuals and families	201	290	715	467	23	21	21	15
Foreign	90	188	531	573	10	13	16	18
Corporations	142	175	415	463	17	12	12	14
Insurance companies	132	200	410	419	15	14	12	13
Endowments and foundations	102	96	267	178	12	7	8	6
	$ 867	$1,423	$3,408	$3,185	100%	100%	100%	100%

Source: Venture Economics, Wellesley Hills, Mass.

reductions. Other Venture Economics data show that pension funds represented only 15 percent of 1978 venture capital flow to private firms, whereas individuals and families provided 32 percent.

Since 58 percent of the flow to independent private venture firms in 1984 came from pension funds, endowments, and foreign sources, why should there be any sensitivity to capital gains tax rates? It is clear that these sources are not concerned about tax rates. Who is concerned, if anyone?

The remaining 42 percent of the investors are obviously influenced by tax policy, and they play a huge role in the earliest stages of growth companies. An enormous amount of seed capital and startup activity is conducted by individual investors and smaller funds which have little or no institutional backing. Most of the larger, institutionally backed funds are highly dependent upon individuals and smaller venture firms to provide them with investment opportunities. During the early 1970s these individuals and funds practically vanished from the scene, and this was one of the major reasons for the sharp decline in venture capital activity. If the people who spend most of their time starting companies stop doing it, then the flow of investment opportunities for other venture investors clearly shrinks. These individuals and small funds have sprung up dramatically since 1978 and account for a sizable amount of the observed new venture activity.

A study published by the Joint Economic Committee (1984) has examined both the kinds of investing and the sources of funding for small, medium, and large venture funds.[1] Smaller funds do indeed focus heavily on seed capital and startup financing and their capital is largely provided by individual investors.

Entrepreneurs, who are the central figures in growth company creation and progress, are also influenced by tax policy. Our experience is that entrepreneurs are typically not interested in or particularly sensitive to details of tax policy, but they are enormously interested in how much of their potential wealth will flow to them after taxes. This subject has also been addressed by Stanley Pratt of Venture Economics. Exhibit 6–1 is a copy of his April 5, 1985 letter to Secretary of the Treasury James Baker.

ARE CAPITAL GAINS TAX RATES IMPORTANT TO THE FLOW OF VENTURE CAPITAL?

The available historical evidence says strongly yes, that capital gains tax rates *are* important to the flow of venture capital. Would it be prudent to ignore that evidence? I think not.

Entrepreneurs, venture capitalists, and other taxed entities play an enormous role at the front end of the process: seed capital and startups. Nontaxed institutions have typically been a much smaller source of capital for this most risky part of venture capital.

Most startups in the United States are not financed by professional venture capitalists, but rather by individuals, families, and friends. Although venture capitalists finance a large number of startups that provide significant growth, job creation, and productivity, other investors play a major role in new company formation.

Capital gains tax rates should be maintained at current levels or, preferably, decreased further.

NVCA REACTION TO THE TREASURY DEPARTMENT'S NOVEMBER 1984 PLAN

The National Venture Capital Association (NVCA) generally considers the Treasury proposal a positive move by attempting to improve the efficiency in which resources are allocated in our country. However, in addition to its fairness aspect, one of the most important goals of this

VENTURE ECONOMICS

April 5, 1985

The Honorable James A. Baker III
Secretary of the Treasury
Department of Treasury
15th and Pennsylvania Avenue NW
Washington, D.C. 20220

Dear Mr. Baker:

There has recently been confusion in the national media as well as in data from the Treasury Department as to the role of non-taxable investors in the funding of the organized venture capital industry. Since Venture Economics, Inc. or our monthly publication, VENTURE CAPITAL JOURNAL, is credited as the information source, it is important that we inform you of the facts.

Tax-exempt (pension, foreign, endowment and foundation) investors are not the principal source of funding for venture capital funds. The confusion was most probably occasioned by analysis of only one of the segments of the venture capital industry, the independent private venture capital firms, but their $11.4 billion represents only 71% of the total industry capital. The other sectors, Small Business Investment Companies ($1.6 billion) and corporate subsidiaries ($3.0 billion) are almost exclusively taxable entities.

Analysis of the complete data shows that the increase in venture capital between 1978 and 1984 was slightly over $13 billion. Tax-exempt investors provided almost $5.4 billion of this total. The current capital committed to the organized venture capital industry is approximately $16 billion of which at least 62% is from taxable sources and only 38% from tax-exempt investors. If one considers informal venture investments from taxable individuals, which some analysts estimate to be greater than the organized activity, tax-exempt investors are even less significant.

Pension fund investors are important to the process since some portion of their more than $1 trillion of investment assets must be attracted to finance new business development. While they will not be directly influenced by tax considerations, they do in fact seek investment opportunities in favorable environments that have been stimulated by tax incentives. Even though many like to credit them with a leadership role, the fiduciary responsibilities of pension, endowment and foundation investors mandate a followers role in new investment trends, after others have served as pioneers.

Potential opportunity, not tax consideration, stimulates institutional investment. Innovators perceiving a favorable environment will create investment opportunities and institutional capital investment will flow towards the proven successes.

It is critical that we understand the real forces that drive new business development -- the perceptions of the entrepreneur. The

Venture Economics, Inc.
16 Laurel Ave., P.O. Box 348, Wellesley Hills, MA 02181
(617) 431-8100 Telex: 94-8637 VENTICON WELL

Page 2

The Honorable James A. Baker

belief that he or she will succeed in building a major new business and significant personal wealth despite the fact that most new businesses fail is a perception that often overcomes stark reality.

The favorable capital gains tax differential not only reinforces this perception, but it is critical in enabling the entrepreneur to attract the key employees who will leave large corporation comforts to work 80-hour weeks to make the vision a reality.

Incentives are relevant because entrepreneurs perceive them to be necessary. How can any of us ignore the sterility of new business development from 1969 through 1977, the period during which there was no differential between capital gains and ordinary income tax rates?

The rationale behind the current Treasury Department tax reform proposal overlooks two important factors. First, increasing the capital gains rate while decreasing the tax rate for dividends and interest income benefits the wealthy at the expense of those seeking to build wealth. In effect, we would hamper the most powerful instrument for social mobility in our economy -- not just for the Steve Jobs' (Apple Computer), but for the many hundreds of middle class employees who joined the effort and shared the success.

The capital gains tax differential can enable young businesses to compete with large corporations for both the capital and, by making Incentive Stock Option plans attractive, the management talent.

Second, our entrepreneurial climate and venture capital infrastructures are key competitive advantages internationally. The Europeans and Japanese recognize this and are attempting to develop a similar environment. Should we take an opposite course and signal entrepreneurial disfavor by eliminating the capital gains tax differential?

There has been discussion recently about providing special treatment for the venture capital industry. I strongly believe that narrowly targeted tax incentives are not feasible in the long term. Entrepreneurs, venture capitalists and investors should be rewarded for success, for achieving growth of jobs and economic benefits, for the creation of new wealth and value, not reimbursed for unsuccessful attempts. The capital gains tax differential rewards risk-taking and patience without undue targeting.

Very truly yours,

Stanley E. Pratt, Chairman
Publisher VENTURE CAPITAL JOURNAL

SEP/

cc: The Honorable Richard G. Darman
Deputy Secretary of the Treasury

The Honorable Ronald H. Pearlman
Assistant Secretary for Tax Policy

proposal should be to stimulate economic growth. In this respect, it falls short in one vital area, as it relates to capital gains. This deficiency will produce a negative economic impact by

- The elimination of the differential between the rate of tax on capital gains and that on other investment income
- An increase in the tax rates on capital gains from 20 percent to 35 percent

These proposals will reduce the availability of funds for new business formation and emerging growth companies. The following are facts.

- Growth companies create jobs and productivity and increase tax revenue.
- Equity capital is the lifeblood of growth companies.
- Equity capital is highly sensitive to the capital gains rate and the differential between the rate of tax on capital gains and that on other investment income.
- The Treasury proposal raises the capital gains rates and eliminates the differential.

Therefore, the Treasury capital gains proposal will

- Lower the rate of job creation
- Lower the growth in tax collections from payroll, social security, and corporate income
- Exacerbate the deficit in the balance of payments
- Reduce the rate of innovation in the economy
- Negatively impact the federal deficit

The Treasury and other flat-tax proposals favor investment in secure income-oriented investments and substantially lower the after-tax returns on high-risk capital gain-oriented investments.

The Treasury proposal will have the following unintended effects:

- Encourage fixed income and high dividend yielding investments
- Penalize high-growth companies whose access to capital is based upon growth instead of dividend yield
- Divert funds from the new issues, over the counter and venture capital markets and disrupt the entire process of economic development and innovation in this country
- Discourage entrepreneurs from leaving secure and high-salaried

positions to risk their careers for the prospect of capital gains

The Treasury proposals penalize investments with high risk/high reward ratios because

- Indexing results in lower taxation of low-growth investments.
- Indexing is a step in the right direction, but if the current capital gains exclusion is eliminated it is inadequate to avoid higher taxation of high-growth investments.

(Assuming a five-year investment with a 5 percent inflation rate, the Treasury proposals result in lower capital gains taxes if the return on investment is 10.5 percent or lower, and if the ROI is higher than 10.5 percent, the proposals result in a greater taxation of capital gains. Note that five-year Treasury bills currently yield approximately 11 percent.)

The tax proposals were intended to be revenue neutral and to simplify the tax code. With respect to capital gains on equities, the tax bill is probably revenue negative and more complex.

The Treasury and other flat-tax proposals that do not maintain low capital gains rates and a differential with other investment income have put other objectives ahead of the health of the economic system. The entire process of new company formation and growth is totally dependent upon the availability of risk capital. Government consumption of the available risk capital pool through high capital gains taxes is self-defeating if the goal is to close the deficit through economic growth.

NOTES

1. Based on the first survey of its kind of the nation's venture capital market, by Robert Premus.

REFERENCES

Birch, David L., and Susan J. MacCracken. 1984. "The Role Played by High Technology Firms in Job Creation." Cambridge, Mass.: MIT Program in Neighborhood and Regional Change.

Horsley and Keogh Associates, "Ten Fund Performance," 1984. (Unpublished.)

U.S. Congress, Joint Economic Committee. 1984. *Venture Capital and Innovation* by Robert Premus. Washington, D.C.: Government Printing Office, December 28.

U.S. General Accounting Office. 1982. "Government-Industry Cooperation Can Enhance the Venture Capital Process." GAO Report to Senator Lloyd Bentsen, Joint Economic Committee, GAO/AFMD−82−35, August 12.

Venture Economics. 1985. "Study of the Economic Impact of Venture Capital-Backed Companies." Wellesley Hills, Mass.: Venture Economics.

7 GUNNING FOR CAPITAL GAINS AGAIN

Oscar S. Pollock

It is hard to exaggerate how much the capital gains tax reductions of recent years have benefited the stock market and aided the raising of risk capital for American business. But the Treasury Department's original tax reform plan would raise maximum tax rates on long-term capital gains substantially. Both common sense and recent historical experience argue strongly against any such change in tax policy.

Decisive shifts in capital gains taxes over the past twenty years provide considerable experience upon which to draw. One lesson is that the trend and level of such taxes on investors have been reflected rather closely in the trend of U.S. equity values. In turn, the level of equity values determines how much common-stock financing can be done.

The new Treasury proposals would subject long-term capital gains to a "one-two" punch. First, the present exclusion of 60 percent for long-term capital gains would be scrapped and all capital gains would be taxed as ordinary income. Then, a provision for indexing the cost basis of capital gains would be added. These steps would raise the maximum rate on long-term capital gains to 35 percent, before indexing.

PART TWO OF THE PUNCH

Part two of the punch occurs because state and local taxes would no longer be deductible for federal income tax purposes. Many states have taxes on capital gains and investment income and they would have a much sharper impact on many taxpayers if deductibility ends. In a number of major states the top tax rates on long-term capital gains would rise to the 40 percent to 50 percent range; for residents of New York City, the top rate could go well over 50 percent. While federal tax rates on short-term capital gains and dividend income would fall, these reductions would be largely offset by the loss of state and local tax deductibility. What effect would such changes have on the equity markets?

Equities are valued on the basis of the current and future returns that their owners perceive them capable of offering, after taxes and after inflation. When tax changes considerably alter the possible tax bite on the components of that return—capital gains and dividends—the changes are reflected in the price structure of equities. Common stocks become worth more or less than they would be otherwise.

Activity in the markets should not be confused with ownership. Institutions now account for most of the activity in the major markets. But most equities are still owned by individual investors. Of the stocks owned by individuals, a large number are owned by those with high incomes, who are very sensitive to tax considerations, particularly capital gains taxes.

Obviously, taxes on capital gains and dividends and changes in these taxes are not the only factors that influence equity values. They combine with other considerations, both economic and psychological, and this makes it hard to measure their impact precisely. But just because tax factors are not alone in influencing equity values and just because their impact cannot be precisely stated does not mean they do not have a profound effect.

Indeed, Figure 7—1 shows that there is a striking long-range correlation between major investor tax trends and the performance of real equity values.

The Treasury tax proposals would push maximum long-term capital gains tax rates back into that range that helped to precipitate a nine-year depression in equity prices; between the end of 1968 and year-end 1977, they lost more than 45 percent of their value in real terms. This prolonged depression greatly reduced opportunities for equity financing and caused a crisis in risk investment. It is easy to forget that this was

Figure 7–1. Total Corporate Equities Market Value, Adjusted for Inflation ($ billions).

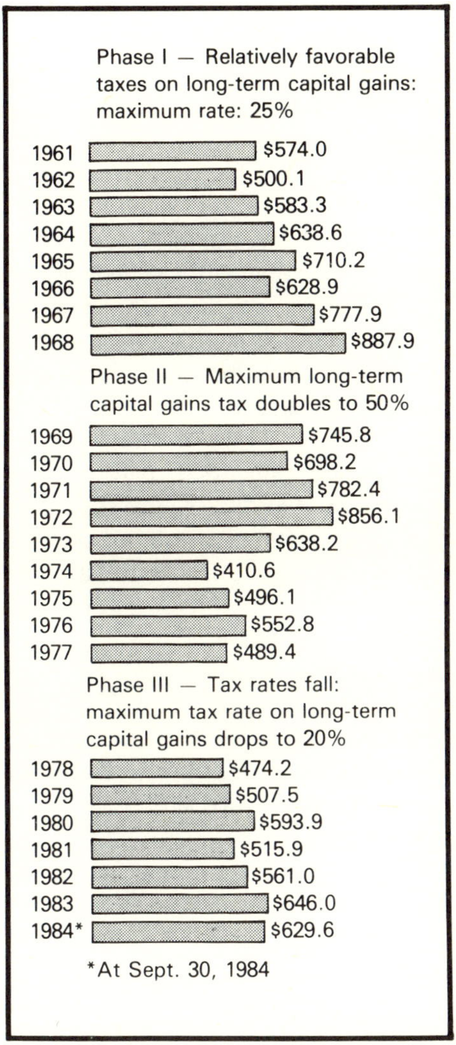

Source: Federal Reserve Bank's Flow of Funds: Total Corporate Equities, adjusted for changes in the *Consumer Price Index,* using 1961 as a base.

a period when venture capital dried up and initial public stock offerings fell to very low levels. In 1979 Paul E. Gray, president of the Massachusetts Institute of Technology, said that high capital gains taxes were one of the principal reasons for America's lagging technological leadership.

The significant 1978 cut in capital gains tax rates started a movement toward lower taxes on individual investors. Since the end of 1977, total equity values have increased again in real terms. Individual risk-taking and entrepreneurial activity have blossomed. In the 1979–1983 period, funds raised by corporations through common-stock offerings averaged $15.2 billion annually, compared with an average of $6.1 billion annually in 1975–1978.

Moreover, at the lower rates, capital gains tax receipts rose in 1979 and 1980 and have substantially exceeded prior projections by the Treasury Department. By comparison, the new Treasury capital gains proposals for individuals and corporations are projected to lose tax revenues, despite some rather generous increments from investors cashing in capital gains during the suggested 1986–1988 transition period.

On September 30, 1984 the total market value of corporate equities was more than double the 1968 total, before adjustment for inflation. But in real terms, the 1984 figure was almost 30 percent below the 1968 year-end amount. It is clear that, in the last sixteen years, much of the gains in equities have been due to inflation, and in this turbulent period indexing the cost basis of capital gains would have been better for many investors than the present exclusion.

Looking ahead, it may be worth adding indexing in some form to protect investors, but this provision should not be substituted for the exclusion. Tax policy should not be based on the premise that the country's economic policy of reducing inflation will fail. If inflation continues to subside, the Treasury proposals would subject the long-term gains of many individuals to maximum tax rates almost twice the current rates, with no provision for greater deductibility of losses. Such changes would stack the odds against investors to a degree that could revive the critical investment problems of the 1969–1977 period.

Indexing does virtually nothing to encourage risk investing. The individual investor is the lead risk-taker and his active participation is essential across a wide spectrum—private business start-ups, venture capital partnerships, initial public offerings, common-stock offerings by small or rapidly growing companies. In such investments, individuals, in effect, buy a number of stocks at $1.00 in the hope that one or two will rise to $20. The possibility of indexing a $1.00 cost basis for inflation would mean virtually nothing to them.

Under inflationary conditions, the indexing provision could increase the investment appeal of large, mature companies that pay handsome dividends. But, quite possibly, the increase would come at the expense of promising growth situations. If investment in such larger, secure companies could produce dividends plus tax-free gains, why should investors take risks—and pay up to 50 percent capital gains tax rates—on the shares of smaller, more speculative companies?

NEW LOCK-IN EFFECT

Indexing threatens to create a new type of lock-in effect; in a period of rising inflation, investors will hold on to their shares in the hope that next year's appreciation will be tax free. Of course, when inflation is low, high tax rates on large, real capital gains would create a lock-in effect of their own. Reduced mobility of capital does the most damage to the valuations of emerging growth companies and the fulfillment of their financing needs.

Investment incentives should be simple and perceptible. While presented as "tax simplification," the Treasury proposals, and particularly all the indexing provisions, would make individual investing much more complicated and its rewards less perceptible because future inflation rates are impossible to predict.

Since the Treasury capital gains proposals could undermine risk investing by individuals, curtail business and job creation and raise the deficit, too, the current provisions—specifically, the 60 percent exclusion for long-term capital gains—should not be changed. One purpose of tax reform should be to build on tax policy successes; it should not repeat the worst mistakes of the past.

8 CAPITAL GAINS TAXATION: INCENTIVE TO GROWTH

Thomas F. Ruhm

The existing capital gains differential creates an enormous incentive to corporate business growth. Stock prices for growth stocks are much higher when there is a substantial capital gains differential (presently the ability to exclude 60 percent of realized long-term capital gains from taxable income) than they are when there is not. The existence of these higher stock prices draws capital into companies that are in the growth mode. The capital comes in both through venture capital private equity investments in start-up and young companies and through underwritten primary public offerings of common stock of all growing companies. (See Tables 8–1 and 8–2 to observe the remarkable growth of venture capital investments and public primary offerings of equity securities after the restoration of the capital gains differential in 1978.)

The *growth mode* is the condition that occurs when a company reinvests all or most of its earnings and profits in its own business and pays out little or no dividends. There is tremendous incentive for growth companies to be in this mode when their significant stockholders would rather obtain the profits from their investments in the form of realized capital gains than in the form of dividends. This occurs when these

Table 8–1. Venture Capital Industry, Estimated Fundings and Disbursements ($ Millions).[a]

Year	Net New Private Capital Committed to Venture Capital Firms[b]		Size of Total Pool	Estimated Disbursements to Portfolio Companies		Public Underwriting of Companies with Net Worth of $5 Mill[illegible] or Less: Number	Amoun[illegible]
1984	$>4,200	↑	$16,300	$3,000	↑	(224)	$ 1,186
1983	4,500		12,100	2,800		(477)	3,671
1982	1,800		7,600	1,800		(113)	619
1981	1,300	$13,400	5,800	1,400	$11,650	(306)	1,760
1980	700		4,500	1,100		(135)	822
1979	300		3,800	1,000		(46)	183
1978	600	↓	3,500	550	↓	(21)	129
			Capital Gains Tax Decrease				
1977	39	↑	2,500–3,000	400	↑	(22)	75
1976	50			300		(29)	145
1975	10			250		(4)	16
1974	57	$ 466		350	$ 2,935	(9)	16
1973	56			450		(69)	166
1972	62			425		(409)	896
1971	95			410		(248)	551
1970	97	↓	2,500–3,000	350	↓	(198)	375
			Capital Gains Tax Increase				
1969	171		2,500–3,000	450		(698)	1,367

Total Estimated Capital Committed to the Organized Venture Capital Industry ($ Billions)

	12/31/82	12/31/83	12/31/84
Independent private venture capital firms	$4.4	$ 8.2	$11.8
Small business investment companies	1.3	1.4	1.6
Corporate subsidiaries (Financial and nonfinancial)	1.9	2.5	2.9
Total	$7.6	$12.1	$16.3

a. This pool remained static from 1969 through 1977 at some $2.5 to $3.0 billion (with new fundings more or less equal to withdrawals).

b. Total new private capital less capital withdrawals.

Source: Venture Economics, Inc., Wellesley Hills, Mass.

significant stockholders are taxpaying individuals whose ordinary income is taxed at a maximum rate of 50 percent and whose capital gains are taxed at a maximum rate of 20 percent. The most significant stockholders in this group, of course, are the top executives and the entire body of employees of the company who have, on a relative basis, substantial equity interests in that company, either directly in shares of stock or indirectly through stock options.

THE NONEFFECT OF NONTAXPAYING ENTITIES

The fact that pension funds may also be substantial stockholders of the company and that pension funds pay no income tax does not at all alter the incentive to plow back earnings, because pension funds do not care what form their investment profits come in. If anything, they may prefer their portfolio companies to plow back earnings to grow themselves. Pension funds want only the highest rate of return compatible with reasonable safety of principal and will take it in the form of either net capital gains or ordinary income. The same is roughly true of foreign investors.

It has been argued that, since pension funds are frequent investors in growing high-technology companies and do not pay taxes, therefore the capital gains tax differential has little to do with bringing investment funds into the high-technology area. Nothing could be further from the case. Many pension funds in the early 1980s invested a prudent percentage of their assets in venture capital investment partnerships and directly in publicly traded growth stocks because those partnerships and those stocks were achieving very high compound annual rates of return at that time. The major reason why this very high rate of return was being achieved was the existence of the 60 percent capital gains exclusion.

The capital gains differential represented by this 60 percent exclusion was restored in 1978. Immediately thereafter the market prices of growth stocks began to climb. A number of venture capital limited partnerships were thereby enabled to take public companies that had just been waiting for the return of the growth stock market. The success of these public offerings imparted strong compound annual rates of return to those partnerships. These rates of return were thereupon exhibited to pension fund managers, who in 1979 were finally given permission under the Employee Retirement Income Security Act

Table 8–2. Statistics on Numbers of Underwritten Registered Primary Public Offerings and Aggregate Dollar Amounts[a] of Such Offerings of Equity Securities[b] from 1976 through 1984, Broken Down by Sizes of Companies.[c]

	All Issues of Equity Securities	
	Number of Issues	Total Amount
1976	268	8,578,974,000
1977	215	6,691,452,000
1978	302	6,154,999,000
1979	377	6,634,993,000
1980	795	17,214,112,000
1981	1083	19,062,588,000
1982	794	18,847,308,000
1983	1839	41,346,685,000
1984	980	13,489,425,000

	Companies with Annual Revenues of $100 Million or Less		Companies with Annual Revenues of $100 Million to $1 Billion		Companies with Annual Revenues of Over $1 Billion	
	Number of Issues	Total Amount	Number of Issues	Total Amount	Number of Issues	Total Amount
1976	121	$ 702,344,000	109	$ 3,623,735,000	38	$ 4,252,895,000
1977	113	684,986,000	74	2,686,590,000	28	3,319,874,000
1978	176	1,140,779,000	108	3,240,776,000	18	1,773,444,000
1979	254	1,196,704,000	91	2,956,951,000	93	2,481,338,000
1980	524	3,568,698,000	205	7,725,812,000	66	5,919,602,000
1981	799	4,955,884,000	188	5,796,283,000	96	8,310,421,000
1982	436	2,934,408,000	165	5,234,960,000	193	10,677,940,000
1983	1237	11,557,001,000	383	13,025,351,000	219	16,764,333,000
1984	779	4,227,722,000	121	3,955,717,000	80	5,305,986,000

a. Aggregate dollar amounts are not adjusted for inflation.

b. Equity securities includes common stock and securities convertible into common stock.

c. The break-downs by size of companies are based on annual revenues.

Source: Securities Data Company, Inc., 62 William Street, New York, New York 10005.

(ERISA) to invest some of their funds in venture capital. This resulted in substantial investments of pension fund money in venture capital limited partnerships in 1979 through 1984.

If the capital gains differential had not existed, growth stocks would not have been especially popular with investors, and pension funds and foreign investors would not have put their money in venture capital investment partnerships or into growth stocks. Without any question, if the capital gains differential were eliminated or substantially reduced, there would be a flight of pension fund and foreign investor capital from venture capital investing.

BACK TO THE INCENTIVE

The incentive to plow back earnings exists for all companies that have individuals as shareholders—and most do. For instance, the mutual fund industry is a heavy investor in corporate equity securities, and profits on mutual fund investments are taxed to mutual fund shareholders in the form they are realized by the fund, for example, as long-term or short-term capital gains or as dividend or interest income.

There is no limit to the size of company that may decide to take on a growth image. This translates immediately into more research and development being done by corporations nationwide and more capital investment when the timing of the business cycle is conducive to it. Starting in 1978 when the capital gains differential was restored (not merely since 1981 when the R&D tax credit was enacted), R&D has swung from an annually decreasing rate of growth to an annually increasing rate of growth—now up to a more healthy approximately 3 percent of gross national product. Furthermore, capital investment in the current business cycle has been cited as being the highest it has ever been in U.S. history.

THE BOTTLING-UP THEORY

To argue against this claim of beneficial macro effects on investment for the 60 percent exclusion of capital gains from taxable income, a number of economists say that keeping investible funds in corporations by imposing high tax rates on dividends and low tax rates on capital gains is economically unsound. They say that it would be better for corporate earnings to be distributed to shareholders, where they can be put to the best economic use, rather than to be reinvested in the corporation's own business, which may not be the best economic use.

There are two answers to this proposition: First, there is almost no restriction today on how a corporation can invest its retained earnings. It can further its own business. It can go into another, related or unrelated, business. About the only thing it cannot do is to buy a competitor or a major customer or supplier. Therefore, encouraging corporations to retain and reinvest their earnings causes 100 percent of their retained earnings to be as well invested as those corporations can conceive of—and there is lots of incentive for corporate management to see to it that that investment produces excellent results (which, according to the free market theory, means that it must have been economically very useful).

On the other hand, if earnings are paid out in dividends, they are immediately subjected to a tax of up to 35 percent (under the President's 1985 tax reform proposals). In many cases, moreover, a state tax is imposed, which can range up to 10 percent (nondeductible under the President's proposals), leaving as little as 55 percent to be reinvested. Then they are further reduced according to the recipients' propensity to consume them. Therefore, probably something less than 50 percent of a corporation's earnings are ultimately reinvested through the dividend route. Obviously, keeping earnings in the corporation is vastly better for national investment levels.

THE OPPOSITION

Certain persons have over the years been extremely desirous of taxing capital gains as ordinary income. These people resist recognizing the extraordinary incentive to long-term investment and capital growth that the capital gains differential presents. The Treasury Department's Office of Tax Analysis, for instance, has a history of accepting only static revenue loss or static revenue gain estimates when new tax measures are being considered and never looking into so-called dynamic revenue loss or dynamic revenue growth estimates. This is consistent with a total refusal by such persons, which is shared by that office, to recognize the incentive effects of the differential.

In November 1984 the Treasury proposed to eliminate the 60 percent exclusion entirely and merely index capital gains for inflation upon realization. The paragraph of the Treasury's memorandum in support of its initial tax proposals that appears under "Risk-Taking" in the section on capital gains taxation, states that capital gains rates will be almost the same under the new regime as they are now and were in the 1950s and 1960s. This is strongly to be disputed when state taxes, which will not be deductible from the federal income tax and will also tax capital gains like ordinary income (see certain New York State proposals), are taken into effect (the effective maximum rate will be a total of 44 percent for residents of New York State, 35 percent federal plus 9 percent state—as against a present 22 percent). It is even more distressing that they did not understand that their proposed equalizing of the tax rate applied to capital gains with those applied to any other forms of investment or salary or wage income would badly damage the market for growth stocks (in fact, rumors, and then publication of their November 1984 proposals, and now continuing attacks on the capital gains differential have severely damaged that market). It is the loss of the differential, through the elimination of all the good effects described, that really hurts growth industries, not the absolute comparison between rates as they are now and rates as they may be after a tax reform bill is passed.

The Treasury's refusal to recognize the power of the differential is also consistent with some economists' long-held desire to eliminate all incentives to move economic progress in one direction or another. I happen to believe that this is an extremely short-sighted, purist policy and that the country will greatly suffer if the differential is removed because of the consequent removal of the incentive to long-term investment in the form of research and development, development of new products and technology and capital expenditures. I think that allowing the U.S. economy to progress in a kind of gray, mushy way without the incentive to build for the future is a mistaken policy. I see this as the major issue that the opposition to the capital gains differential presents.

On the other hand, the Treasury's supporting memorandum of November 1974 recognizes that

> taxing real (indexed) capital gains at a maximum ordinary income rate of 35% will result in a greater tax burden on the most successful investments made by venture capitalists. If one assumes sufficiently high rates of return and moderate rates of inflation, indexing for inflation, even over the approximately seven to ten year life of the average venture capital investment, will not be as generous as the 60% exclusion.

The problem with this is that it is the results of those most successful investments that most greatly benefit our nation! We badly need a strong incentive to create those successes! This strong incentive is presently provided by the effects of the capital gains differential upon the value of entrepreneurs' and their associates' stock interests in the companies they are creating. The capital becomes available for the growth of their companies, and the market wants the stock they have to sell. This is the goose that lays the golden eggs.

The empirical evidence to support our side of the argument is how beautifully growth industries have done for the last six years since the enactment of the Steiger bill (see the data on Tables 8−1 and 8−2, for instance) and how well they did in the 1950s and 1960s, when a strong differential also existed. The opposition has no empirical evidence for their side.

THE BENEFICIARIES OF LOW CAPITAL GAINS RATES

Who benefits from lower capital gains tax rates? We estimate in my company that 90 percent of the capital that we invest directly in small growing companies goes to pay the salaries of those companies' personnel. They in turn develop the technology that becomes a marketable, profitable product. We see to it that not only management but all employees of companies in which we invest have substantial amounts of stock or stock options in the companies that employ them. Only in this way can our private equity investments become profitable with the proper incentive to grow themselves without management interference from investors.

Large companies also benefit substantially from lower capital gains rates. Principally the benefit comes from their being able, with stockholder approval, to retain greater amounts of their annual earnings to finance increased research, product development, and other capital investment. They do this in order to enhance the growth quality of their stock and just because it is in the nature of people to want to increase their companies' sizes. The data on Table 8−1 shows how much easier it became for large companies, among others, to raise equity capital in underwritten public offerings after the capital gains differential was restored in 1978. Large companies also benefit from the prodding competition from growing, smaller companies. This gets them to change their technologies and improve their products faster, giving them an improved public image and better employee morale.

Benefit from lower capital gains taxes does not stop with entrepreneurs, employees, companies and investors. Society at large is a substantial beneficiary also, for the following reasons: First, the new products developed by these growth companies bring about a better standard of living for everybody. Second, if it were not for the new technology developed by growth companies, the larger, established companies would be much slower to shift to newer and more efficient processes and products. Two examples of this are, first, IBM's hesitancy to shift from core memory to silicon chip memory devices because of its huge investment in core memory manufacturing facilities and, second, AT&T's reluctance to develop a packet-switching network for data transmission because of the high profitability to them of leasing entire lines, usable by only one party at a time, to companies or persons conducting data transmissions. In both cases, venture capital and public capital invested in existing growth companies and new young companies developed the highly efficient alternative technologies of silicon transistor memory chips and packet-switching networks for data transmission. Without the capital gains tax differential there would be a substantial dampening effect upon the present degree of technological development in this country.

Society at large further benefits from the capital gains tax differential through the availability to its imaginative, energetic young people of jobs with growing companies of all sizes as opposed to jobs only with large or giant companies in which the individual from time to time gets lost completely. Many people work much more efficiently and more happily in a small or medium-size employment situation. The work atmosphere tends to be more dynamic and flexible, especially when each working person has a meaningful "piece of the rock."

We also benefit from increases in both jobs created and taxes paid (both by corporations and their employees) resulting from the formation of new businesses and the growth of existing businesses.

Furthermore, the growth of small companies does not in any way impede the growth of the large companies. Quite frequently, when a small company has grown to a substantial size, such as $100 million in sales, it becomes an attractive candidate for acquisition by a larger company and subsequently becomes a division of that company.

IS THERE ANY OTHER WAY TO PRODUCE EFFICIENT GROWTH?

Could straight government subsidies to business achieve the same growth as the capital gains tax differential? My answer to that is an emphatic "No." Capital gains do not spring automatically from well-intended efforts at creating new technology. They come only where the investment and entrepreneurial efforts have been intelligently directed at a technology that is needed in the large economic marketplace. It takes the professional venture capitalists and underwriters to ascertain which technologies have that potential and which management teams have the necessary qualities of good management to realize that potential. These professionals are given the necessary incentives to do their jobs right by competition from other investing groups and by the profit motive. Just spraying government money at small companies will not achieve the quality of results achieved by professional investors.

People who have something to gain from the correctness of their decisions will make better decisions than those who are simply drawing a regular salary. Venture capitalists are generally compensated with a substantial share of the capital gains realized from the investments that they make for their firms. The typical split is 80 percent for the limited investors and 20 percent for the working venture capitalists. That gives the venture capitalists the necessary incentive to find profitable enterprises in which to invest the firm's money. If they do not find such enterprises, there are no capital gains to share in.

By the same token entrepreneurs are enticed away from the large companies, for whom they typically work prior to setting up their own companies, by the prospects of substantial capital gains to be realized upon the ultimate sale of the stock that they receive in their new companies. But if they fail to make the company grow, they will not receive any of this substantial incentive compensation. Therefore, there is a strong inducement to the entrepreneurs to develop their technology successfully. Take away these incentives and you will surely take away much of the rapid technological progress that our country is presently making.

MORE BENEFITS

One remaining area of benefits to society from the capital gains differential, but by no means the least important, is productivity. Productivity is the measure of the rate of output by our country's workers. It was interesting that in the GAO (1982) study of venture capital, in which they studied particularly the seventy-two companies backed by venture capital that had gone public during the period from 1970 through 1979, they found that every one of the products produced by these companies enhanced the productivity of either white collar or blue collar workers. It stands to reason that products that enhance productivity find strong markets awaiting them and hence are popular subjects for entrepreneurial efforts.

Finally, the United States' worldwide reputation for high-technology development has been vastly enhanced over the past seven years since the passage of the capital gains tax reduction in 1978. This contributes to U.S. national security and makes American products more exportable, thus enhancing our economic position in the world.

CONCLUSION

Making the rate of tax on long-term capital gains equal to or approach that imposed on ordinary income ipso facto causes the loss or substantial reduction of the capital gains differential. If that happens, there will be much less incentive for companies to plow back profits to do more R&D and capital investment. There will be vastly fewer investments in new small companies or existing growth companies generally. The business world will be dominated by sluggish large companies. If that is the goal, advocate making the capital gains tax approach or equal the tax rate on ordinary income. To avoid that fate, advocate keeping the 60 percent deduction for realized capital gains—or even raising the deduction to 100 percent (a zero rate of tax on long-term capital gains).

9 ACRS-ITC, CCRS, AND NCRS COST RECOVERY IN AN INFLATIONARY ECONOMY

Norman B. Ture and Carlos E. Bonilla

The appropriate design of the capital cost recovery provision in the income tax is a long-standing tax policy issue. The dimensions of this issue have come into clear focus since the introduction of the Accelerated Cost Recovery System (ACRS) and the liberalization of the Investment Tax Credit (ITC) in the Economic Recovery Tax Act (ERTA) of 1981. The ACRS-ITC has been criticized as affording a tax subsidy for capital formation in general and for differentially reducing the effective tax rates applicable to the income produced by different kinds of capital. Fundamentally at issue are the requirements that the concern for tax neutrality impose on the specifications of a capital cost recovery system.

TAX NEUTRALITY

By tax neutrality, we mean that a tax does not alter the relations among prices and costs that would prevail in the absence of taxes. Suppose, for example, that without any taxes the price of a pound of apples is half that of a dozen oranges, but when a particular tax is imposed, the prices are the same. Obviously, the tax has increased the price of apples relative to that of oranges; except under extraordinary conditions, this relative price change will lead people to buy fewer apples compared with oranges than they would have if the tax had not been imposed.

THE BASIC INCOME TAX DISTORTION

In this sense, the income tax is nonneutral with respect to consumption, on the one hand, and saving and investment, on the other. It raises the cost of future income compared with that of current consumption. To see this, compare these relative costs in the absence and presence of an income tax.

To begin with, consider a world with no taxes where $1.00 can either be used to buy an asset which produces a perpetual income stream of 10 cents a year or can be used to purchase $1.00 of current consumption. If we now impose an income tax at, say, 50 percent, we alter the relative prices of consumption and future income. The dollar of pre-tax income now buys 50 cents of consumption goods. That same dollar now can buy an income stream of 5 cents, which itself is taxed, leaving a net income stream of 2.5 cents. The tax has doubled the number of pre-tax dollars needed to maintain the level of consumption but quadrupled the number of pre-tax dollars needed to purchase an equivalent stream of future income. It has, therefore, doubled the cost of future income relative to the cost of current consumption. The higher the tax, the greater the distortion in favor of current consumption.

INCOME TAX DISTORTION OF CAPITAL ALLOCATION

We can think of this fundamental bias of income taxes against saving and investment and in favor of consumption as a first-level nonneutrality. In addition, insofar as it allows differing deductions, credits, and exclusions, the income tax is likely to alter the relative costs of acquiring or holding different kinds of capital. These divergent costs of capital are likely to result in a different allocation of saving among alternative capital uses and a different composition of capital than if there were no taxes. These income tax induced differences in capital costs are conveniently designated second-level nonneutralities.

Both first- and second-level nonneutralities clearly impair the economy's efficiency. The first-level bias, by distorting saving-versus-consumption choices, results in a smaller stock of capital, less production capability, and a lower standard of living than would prevail if the tax were more nearly neutral. And if we can fairly assume that the way people allocate their saving and investment in the absence of taxes would result in the most productive stock of capital, then the second-level distortions must also impose significant costs on the economy. Of the two, it should be clear, the first-level nonneutrality, the basic income-tax bias against saving and investment in general, must be the more costly.[1]

NEUTRAL TAX TREATMENT

Two alternative means to eliminate the basic bias against saving and investment are obvious from the foregoing example: either exempt saving (that is, capital outlays), from taxation while taxing the gross returns to capital, or tax the current income which is saved and invested and exempt all the returns to capital.

If current saving is excluded from the tax base ("expensed"), the dollar of current income that is saved and invested escapes the tax; the returns on the investment, however, are fully taxed. In the example, the saved dollar produces 10 cents of income per year before tax and 5 cents a year after tax. In the other approach, the 50 cent tax on the dollar of current income leaves 50 cents to save and invest, producing 5 cents of income a year before tax; because no tax would be imposed on this return, it also provides 5 cents of income a year after tax. In either case, a dollar of current income that is saved bears the same burden as a dollar of income that is consumed; the trade-off between current consumption and future income is the same as if there were no tax.

Either approach, consistently applied, would be neutral in the second-level sense as well as with respect to the overall cost of saving relative to consumption. Indeed, either approach is virtually essential if neutrality of tax treatment among alternative forms of saving and investment is to be assured.

WAS ACRS-ITC TOO GENEROUS?

The ACRS provisions and the liberalization of the ITC enacted in the Economic Recovery Tax Act of 1981 aimed at reversing the slowdown in the growth of the U.S. industrial base. For the five years ending in 1979, increases in real net business fixed capital averaged just over 2 percent of net national product, one-half the rate for the late 1960s. Numerous factors contributed to that slowdown, chief among which was the escalating cost of capital attributable to the basic income tax bias against saving and capital formation, the inadequacy of depreciation allowances to offset that bias, and the intensification of that tax penalty by inflation. High inflation had caused a large discrepancy between the historic and current replacement costs of investment goods. In real terms, the value of the total depreciation deductions allowable with respect to a given amount of depreciable facilities was a fraction of the expenditures incurred for those assets. As a consequence, real marginal tax rates on the net returns on depreciable property were substantially higher than the statutory rates. This in turn elevated the cost of committing saving to depreciable capital.

ERTA made three major changes in cost recovery to address this problem.[2] For one thing, ACRS discarded the notion of useful life and instead established four cost recovery periods of three, five, ten, and fifteen years, depending on the type of property, with the fifteen-year category reserved for all real property not designated as ten-year class property. In general, capital cost recovery allowances, were to be computed by use of the 150 percent declining-balance method for property placed in service in the years 1981 through 1984, 175 percent declining balance for property placed in service in 1985, and 200 percent thereafter.

Second, the ITC rate was increased for most eligible property. Under prior law, property with a useful life of less than three years obtained no ITC, that with useful lives of three or four years got a credit of 3.5 percent, that with lives of five and six years got a 6.66 percent credit, while property with a useful life of seven or more years got a 10 percent ITC. Under ERTA, three-year property got a 6 percent ITC and all other eligible property was given a 10 percent credit. ERTA also extended eligibility for the ITC to certain kinds of property formerly excluded from the credit's application. In addition, it extended the period of years over which unused credits could be carried forward before being lost.

The third major feature of the 1981 capital recovery provisions was the safe-harbor leasing provision. Safe-harbor leasing created a mechanism by which firms that were currently unable to use their investment credits and ACRS deductions could transfer these rights to firms that could take advantage of them. It signaled recognition of the fact that the effectiveness of ACRS and the liberalized ITC in offsetting the income-tax bias against investment in durable production facilities depended on the extent to which taxpayers could actually utilize these provisions currently, rather than having to defer their benefits, possibly for extended periods of time.

The Tax Equity and Fiscal Responsibility Act of 1982 (TEFRA) made significant and serious changes in the capital cost recovery provisions. Among its other features, TEFRA instituted partial basis adjustment for the ITC, stopped the phase-in toward the 200 percent declining balance recovery method, limiting the recovery rate to the initial 150 percent rate, and eliminated the safe-harbor leasing provisions. These changes occurred amid charges that the recovery provisions were too generous and that they represented an investment subsidy.

Use of the neutrality concept defined previously makes it possible to test the claim that ACRS-ITC offered excessively generous capital cost recovery. In its simplest form, the test is whether the sum of the present values of the ACRS deductions plus a deduction, equivalent in its effects on tax liability to the ITC, equals, exceeds, or falls short of the amount of the investment. If the sum of the present values of these capital recovery deductions just equals the amount of the investment in the property on which they are allowed, the capital recovery deductions afford neutral tax treatment of that use of saving. If the sum of the present values is less than the investment, these deductions do not fully offset or neutralize the tax bias against that investment. If the sum of the present values exceeds the investment, the capital recovery provisions subsidize the investment.

These present values, at the recovery rates applicable during the ERTA phase-in (150 percent declining balance for assets placed in service in 1981–1984, 175 percent declining balance for 1985, and 200 percent declining balance for 1986 and after), are presented in Table 9–1. Three significant qualifications must be noted before assessing the ACRS-ITC provisions against the tax-neutrality standards. Taking account of these markedly affects the interpretation of Table 9–1.

For one thing, the present value of the ACRS deductions and the ITC must be determined with respect to the time at which the costs of acquiring the property are incurred. For the most part, the ACRS

Table 9–1. Present Values of ACRS-ITC Deductions[a] per Dollar of Investment, with a 4 Percent Real Rate of Discount, at a 5 Percent Inflation Rate.[b]

Recovery Period (Year)	ERTA Provisions Placed in Service in 1981–1984	1985	1986+	TEFRA Provisions (All Years)
3	1.00	1.01	1.01	.97
5	1.01	1.04	1.04	.97
10	.89	.94	.94	.85
15 (real property)	.60	.60	.60	.55[c]

a. The ITC-equivalent deduction is computed as the ITC rate divided by the top marginal corporation income tax rate. The ITC rate is 6 percent for three-year property and 10 percent for five- and ten-year property.

b. At a zero inflation the present values are greater than 1 for all except real property.

c. Recovered over eighteen years.

deductions are first allowed for the taxable year in which the property is placed in service; the ITC is also claimed in the one taxable year. That is often a later taxable year than that in which the investment, or a part of it, is made. This is particularly likely to be true in the case of investment in properties which have an extended, multiyear construction or manufacture period, during which the investor makes progress payments to the supplier of the capital facility. When this is the case, the ITC-equivalent deduction and the ACRS allowances must be discounted to the time (or times) at which the investment cost is incurred to determine whether the present value of these deductions equals, exceeds, or falls short of the amount of the investment.

A second condition is whether the ACRS deductions and ITC are allowed in determining all of the taxes which are imposed on the returns on the investment. There are practically no situations in which this condition is satisfied. When a corporation acquires depreciable property, for example, the ACRS deductions and the ITC are allowed only in determining the corporation income-tax liability. Some or all of the income produced by the property is almost invariably subject to the individual income tax, to the extent the corporation pays dividends out of this income or the individual shareholder sells stock in the company and realizes a capital gain reflecting the retention of part or all of the earnings produced by the property. The income in the hands of the corporation is also likely to be subject to state income tax; few state income taxes have capital recovery provisions which are the same as the ACRS and ITC. Insofar as dividends are paid from these earnings, moreover, the individual shareholders are also likely to be subject to state income taxes. Property taxes also apply; they may be treated as income taxes imposed on the capitalized value of the net earnings produced by the property. In short, the ACRS and ITC are taken into account in determining only a fraction of the taxes imposed on the income produced by eligible capital. This consideration must be given full weight in determining whether these capital recovery provisions subsidize investment.

The final condition is whether the ACRS and ITC are given full effect in the years in which they are claimed or whether their effect on tax liability is deferred until a later taxable year or years. This deferral effect occurs when taxable income before ACRS deductions is less than the amount of the ACRS deduction allowable in that year or when tax liability before the ITC is less than the amount of the credit. In computing the present value of the ACRS deductions and the ITC, therefore, only the respective amounts actually having effect on tax liability in each year should be taken into account. Table 9—1 does not reflect any of these three conditions. It overstates accordingly, the extent to which the ACRS-ITC provisions satisfy the neutrality criterion.

Looking at the ERTA provisions of Table 9—1, it can be seen that the 1981—1984 provisions with a moderate inflation rate (5 percent) closely approximated full cost recovery for three-year and five-year property. The 1985 and 1986 phase-in would have brought three-year property slightly above full recovery; the present value of five-year property allowance would have been about $1.04 per dollar of investment. Before and after phase-in allowances on ten and fifteen-year property would have remained substantially short of full recovery. For these assets, ACRS-ITC as provided in ERTA could not result in a subsidy. These provisions might be perceived as affording subsidies to three- and five-year assets if one disregards the three factors identified above which qualify the evaluation of capital recovery against the neutrality standards. It is useful at this point to frame the discussion in terms of effective tax rates, not only in present value terms.

The effective tax rate on the returns on investment is properly defined as the percentage difference between (1) the present value of the pre-tax gross returns required if the present value of the after-tax gross returns are to just equal the the amount of the investment, and

(2) the present value of the gross returns that would be required if there were no tax. The effective tax rate depends on the present value of cost recovery allowances per dollar of capital outlay and the statutory tax rates applicable to the difference between gross income and capital recovery allowances.

Table 9—2 presents the effective tax rates, under the original ERTA provisions, on the income of assets placed in service in 1986, assuming other taxes are imposed at various rates. The first column is the effective tax rate that would exist if no taxes other than the federal corporate income tax were imposed. On the other hand, if state and local taxes at a rate as low as 50 percent bear on the gross returns produced by these assets; all of them face nonnegative, highly positive, effective tax rates.

For example, if other taxes are ignored and with 5 percent inflation, the effective tax rate on five-year property under ERTA, as shown in Table 9—2, is −3.7 percent when the real after-tax discount rate is 4 percent. If other taxes amount only to 5 percent, the effective rate on this property becomes highly positive, 6.1 percent. With other taxes aggregating 10 percent, the effective rate becomes 18.1 percent.

For long-lived property, the effect is even more pronounced. For example, in the case of fifteen-year real property, for which no ITC is available, the effective tax rate is 48.3 percent when other taxes are 5 percent, compared with 34.6 percent when no other taxes are accounted for. At higher discount factors, such as when the inflation is higher than 5 percent, the impact of other taxes is even more severe.

A major implication of these results is that even with the ITC, the ERTA capital recovery provisions fell short, almost across the board, of eliminating the tax bias against investment in depreciable property, because of the impact of other taxes on the net returns on such property. Stretching out recovery periods or reducing the ITC increases this adverse impact, particularly for long-lived property and even at moderate inflation rates.

The effect of TEFRA was to lower the present value of cost recovery benefits and to raise effective tax rates compared with those that were in place in the first phase of ERTA. As can be seen in column 4 of Table 9—1, the present value of ACRS/ITC deductions made TEFRA benefits fall far short of full cost recovery, even without allowing for the factors identified that bias these results upward. Moreover, the repeal of safe-harbor leasing brought on by TEFRA increases the importance of the last of the three qualifications: the present values, and hence effective tax rates, under TEFRA implicitly assume the

Table 9–2. Effective Tax Rates with a 5 Percent Inflation Rate and 4 Percent Real After-Tax Discount Rate at Selected Rates of Other Taxes.

ERTA Provisions (1986 Plus)

Property Class	Rate of Other Taxes 0%	5%	10%
3	−1.0	9.1	21.5
5	−3.7	6.1	18.1
10	−5.2	15.9	29.1
15 (real)	34.6	48.3	65.2

TEFRA Provisions

Class	Rate of Other Taxes 0%	5%	10%
3	2.7	13.2	26.1
5	2.7	13.2	26.0
10	12.7	24.2	38.3
18 (real)	38.7	52.9	70.2

availability of other income adequate to absorb currently all of the cost recovery. If that is not the case, the deferral of these cost recovery allowances will further reduce their present value.

DOES ACRS-ITC DISTORT INVESTMENT DECISIONS?

ACRS-ITC, either as specified in ERTA or as amended by TEFRA, does not fully meet the tests of tax neutrality. Compared to the capital cost recovery system under pre-ERTA law—the so-called Asset Depreciation Range (ADR), however, the ERTA provisions represented a major advance with respect to both first and second level neutrality criteria. Although ACRS-ITC may distort investment decisions compared with a perfectly neutral cost recovery system, the current law's provisions contribute to more efficient capital allocation than did the prior law's ADR.

The present value of the sum of ADR allowances and the ITC-equivalent deduction per dollar of investment in properties with various ADR midpoint lives is shown in Table 9—3. The ADR lives shown in the table are representative of the kinds of production facilities for which ACRS provided the corresponding, much shorter recovery periods presented in the second column of the table.

Comparing the results in Table 9—3 with columns 1—3 of Table 9—1 shows that the ACRS-ITC provisions significantly reduced the spread in the present value of capital recovery allowance which occurred under ADR-ITC. With a given tax rate, the tighter is the clustering of present value of capital recovery deductions, the smaller is the range of effective tax rates. Comparing the effective tax rates with ADR shown in Table 9—4 with those in Table 9—2, column 1, it is clear that ACRS-ITC provisions significantly reduced the dispersion in effective tax rates. For the same reason, ACRS-ITC must have mitigated, not aggravated, tax-induced distortions of investment.

Moreover, comparison of ACRS-ITC with ADR demonstrates that in addition to reducing the interasset distortion ACRS-ITC also reduced the bias against all investment in depreciable property. ACRS-ITC, therefore, should be seen as contributing significantly to reducing both the dimensions of the tax nonneutralities of prior law.

The same cannot be said for the TEFRA modifications to ACRS-ITC. The range of present values of cost recovery allowances under TEFRA (Table 9—1, column 4) is greater than under any stage of original ERTA. TEFRA, therefore, produces more interasset distortion, accompanied in all cases by a greater bias against investment.

ECONOMIC DEPRECIATION AND TAX NEUTRALITY

Economic depreciation has long been advocated by its proponents as the only neutral method of cost recovery. This view received considerable support when the U.S. Treasury Department (1984: 154—55) in its tax reform proposal stated that tax liabilities should be imposed on the basis of real economic income:

> In the case of investment in depreciable property, measurement of real economic income requires an allowance for the property's economic depreciation. If that allowance is understated, income from the investment is overtaxed and a tax disincentive is created which impairs capital formation and retards the economy's productive capacity. By the same token, overstating depreciation and thus understating income creates an artificial incentive for

Table 9–3. Present Value of ADR[a] and ITC-Equivalent[b] Deductions per Dollar of Investment and Effective Tax Rates, by Selected ADR Midpoint Lives (and Correspondence ACRS Recovery Periods) at 5 Percent Inflation Rate and 4 Percent Real After-Tax Discount Rate.

ADR Midpoint Life (Years)	Corresponding ACRS Recovery Period	Present Value of ADR-ITC Allowances	Effective Tax Rates
3	3	$.93	3.7%
10	5	.92	7.2
18	10	.78	19.1
35 (real property)	18	.41	50.6

a. ADR deductions computed with the 200 percent declining balance method, switching to straight line; the half-year convention is used throughout.

b. The ITC-equivalent deduction is computed as the ITC rate divided by .46, the top marginal corporation income tax rate. The ITC rate for three-year property is 3.33 percent, that for the other property shown (except for real property) is 10 percent.

> one form of investment over another, discriminates among companies within an industry, and encourages nonproductive, tax-motivated investment activities.
>
> The proper measure of economic depreciation in any year is the amount of decline in the real value of an asset over the year, which is equal to the cost of replacing the lost productive value. Due to inflationary increases in replacement costs, pre-ACRS depreciation deductions for many assets understated actual economic depreciation and thus resulted in overtaxation of the income from such assets.

Two distinct themes run throughout these two paragraphs, and both must be dealt with to resolve the issue of economic depreciation. The first of these is that only economic depreciation is neutral, while the second implies that one would only wish to accelerate capital recovery to counteract the effect of inflation.

Economic depreciation, even without the complication of inflation, is in serious conflict with neutrality. While there is no denying that it penalizes capital investments to extend the cost recovery period beyond the life of an asset, it does not necessarily follow that accurately measuring the asset life removes this penalty. If recovery periods were at any time greater than the life of the asset, then a move towards economic depreciation would, in all likelihood,[3] serve to ameliorate the

Table 9–4. Present Value of Cost Recovery Allowances per Dollar of Capital Outlay under ACRS,[a] CCRS, and NCRS at Selected Real After-Tax Rates of Return.

A. Inflation Rate: 0 percent

Recovery Class ACRS (Years)	CCRS	NCRS (Years)	Present Value of Allowances, Real After-Tax Rate of Return: 3.5 Percent ACRS	3.5 Percent CCRS	3.5 Percent NCRS	4.0 Percent ACRS	4.0 Percent CCRS	4.0 Percent NCRS
3	1	4	$1.05	$.94	$.98	$1.04	$.93	$.97
5	2	6	1.08	.93	.99	1.07	.92	.97
5	3	6	1.08	.91	.99	1.07	.90	.97
5	4	6	1.08	.87	.99	1.07	.87	.97
10	5	15	1.02	.85	.98	1.01	.84	.95
18	6	25	.77	.64	.99	.75	.61	.94

B. Inflation Rate: 5 Percent

Recovery Class ACRS (Years)	CCRS	NCRS (Years)	Present Value of Allowances, Real After-Tax Rate of Return: 3.5 Percent ACRS	3.5 Percent CCRS	3.5 Percent NCRS	4.0 Percent ACRS	4.0 Percent CCRS	4.0 Percent NCRS
3	1	4	$.97	$.92	$.96	$.97	$.91	$.95
5	2	6	.97	.91	.96	.97	.90	.95
5	3	6	.97	.89	.96	.97	.88	.95
5	4	6	.97	.86	.96	.97	.85	.95
10	5	15	.85	.83	.96	.85	.82	.93
18	6	25	.55	.63	.96	.55	.60	.91

C. Inflation Rate: 10 Percent

Recovery Class ACRS (Years)	CCRS	NCRS (Years)	Present Value of Allowances, Real After-Tax Rate of Return: 3.5 Percent ACRS	3.5 Percent CCRS	3.5 Percent NCRS	4.0 Percent ACRS	4.0 Percent CCRS	4.0 Percent NCRS
3	1	4	$.91	$.90	$.94	$.91	$.89	$.93
5	2	6	.88	.89	.94	.88	.88	.93
5	3	6	.88	.87	.94	.88	.86	.93
5	4	6	.88	.84	.94	.88	.83	.93
10	5	15	.74	.81	.94	.74	.80	.91
18	6	25	.43	.63	.94	.43	.60	.89

a. ACRS allowances, except for eighteen-year property, include the deduction equivalent of ITC, at a 46 percent tax rate.

distortion. But lengthening cost recovery periods to some estimate of economic life will accentuate the distortion. Economic depreciation, unlike the question of disortion from ACRS-ITC, is an absolute issue, not a relative one.

The problem with economic depreciation is not a pragmatic issue of determining economic lives and actual depreciation patterns, as is so often argued. Rather, the fault is economic depreciation would inevitably violate neutrality, regardless of how well this economic depreciation is measured. Economic depreciation can be neutral neither among investments nor with respect to investment on the whole.[4]

THE EFFECT OF INFLATION ON COST RECOVERY

The recent emphasis placed on inflation indexing for capital recovery has obscured the real issues in the design of capital cost recovery systems. As important as it may be for insulating cost recovery allowances against erosion by inflation, indexing cannot in and of itself transform an inadequate cost recovery system into an optimal one. Indexing can maintain the desirability of an already desirable system, its absence can make an otherwise efficient system fail upon implementation, but it can do no more than prevent a flawed system from becoming more undesirable. Even this last result can only be achieved if the indexing accurately accounts for the effect of inflation. Indexing of capital recovery, moreover, is not the only way to adjust for the effects of inflation.

With short recovery periods and where recovery allowances are greatest in the early years, a capital cost recovery system is less affected by inflation than a system with longer recovery periods. The absence of indexing in such a system may be, therefore, less of a liability than one might imagine.

The real issue concerns the adequacy of a capital recovery system in neutralizing the antisaving, anti-capital-formation bias inherent in income taxation. To be sure, indexing may well be a highly useful element of a capital recovery system in an inflationary environment. In itself, however, indexing does not address the main issue.

Two indexed recovery systems illustrate this point. The Capital Cost Recovery System (CCRS) put forward in President Reagan's tax proposal and the Neutral Cost Recovery System (NCRS) advanced by the Kemp-Kasten "Fair and Simple Tax Act" as introduced in the Ninety-ninth Congress both include indexing. While NCRS would be a major advance toward neutral tax treatment of investment in depreciable facilities, CCRS would be a major retreat.

The CCRS structure is a modification of the Real Cost Recovery System (RCRS) outlined in the original Treasury tax proposal "Tax Reform for Fairness, Simplicity, and Economic Growth." The RCRS structure was obstensibly built on the concept of economic depreciation, that is, the loss in the market value of the asset between the beginning and end of a period.[5] The CCRS structure that evolved from RCRS has six recovery classes opposed to seven in RCRS, while the recovery periods range from four to twenty-eight years compared to five to sixty-three years under RCRS.[6]

NCRS, on the other hand, is a modification of the current law, Accelerated Cost Recovery System. It extends the ACRS recovery periods from the present three to eighteen years to a range of four to twenty-five years. NCRS also increases the nominal amount of recovery allowances so that, before indexing, their undiscounted amounts exceed 100 percent of the expenditure on the asset. For each recovery period, the increase is such that when discounted at a real discount rate of 3.5 percent, the present value of the allowances just equals 100 percent of the expenditure on the asset. Both CCRS and NCRS repeal the Investment Tax Credit.

Table 9—4A presents the present value of cost recovery allowances, by asset class, in a zero inflation world under ACRS (with ITC), CCRS, and NCRS. The results are shown at both a 3.5 percent and 4 percent after-tax real rate of return, the rates assumed in the Kemp-Kasten and administration proposals, respectively.

The values with zero inflation shown in Table 9—4A provide an important point of reference. They show the extent to which the basic feature of the respective cost recovery systems result in distortions, free from the added distortions resulting from inflation. Indexing cannot reduce the distortions that are shown in the zero-inflation case, it can only moderate the additional inflation-induced distortions.

The table shows very clearly that the administration's proposed CCRS falls far short of achieving neutrality. The present values of these allowances fall steadily and steeply as the recovery period is increased and are substantially below full cost recovery for all assets. Indeed, the present value of the class 6 (or twenty-eight year) cost recovery allowance is only slightly more than two-thirds the value of a class 1 (or four-year) asset.

In sharp contrast, the present law ACRS with ITC shows a much closer clustering of the present values of cost recovery allowances except for the eighteen-year property for which no ITC is available. The NCRS allowances afford virtually complete neutrality of both types when the allowances are discounted at a 3.5 percent rate, the real after-tax rate of return used in the determination of the recovery amounts per year for each asset class in the Kemp-Kasten proposal. Discounting at 4 percent, NCRS (using the same amounts per year for each property class as designated in the case with 3.5 percent discounting) falls a bit farther shy of perfect neutrality, but even so it far more closely approaches this goal than does CCRS. ACRS with ITC compares quite well, except in the case of eighteen-year property. Clearly, either NCRS or ACRS with ITC much more nearly satisfies the interasset neutrality criterion than CCRS. This is not the result one would have expected given the emphasis Treasury claims to have placed on neutrality.

When inflation is taken into account, the results with regard to interasset neutrality are much the same as those found when the inflation rate is assumed to be zero. As Table 9–4B shows, NCRS, with its indexing, affords the same present value of its allowances for all classes of property when inflation is assumed to be at an annual rate of 5 percent and the real after-tax rate of return is 3.5 percent.

With 5 percent inflation and a 4 percent real after-tax rate of return, NCRS results in a small disparity in the present value of its allowances in the case of fifteen-, twenty-, and twenty-five-year recovery property. The same sort of results are found when the inflation rate is assumed to be 10 percent a year, as shown in Table 9–4B. CCRS, on the other hand, shows the same substantial decreases in the present values of cost recovery allowances as the recovery periods are lengthened that are seen in Table 9–4A. ACRS, lacking indexing, fares less well under inflation than with a zero inflation rate in terms of the inter-asset neutrality goal. It remains, nonetheless, superior to CCRS in this respect at a 5 percent inflation rate (except for ACRS class 18 assets which do not receive an ITC) although inferior to it at the 10 percent inflation rate.[7]

In formulating economic policy, one certainly should not assume that inflation has vanished from American economic life and that a return to 10 percent-plus inflation is out of the question. Nevertheless, it is ludicrous to devise a system such as CCRS that is superior to current practice only under the assumption of oppressively high rates of inflation. NCRS, in contrast with CCRS, does not rely on this assumption; it much more effectively copes with the twin problems of inflation and neutral treatment among different kinds of capital goods than does either CCRS or ACRS. NCRS fails to reach perfect cost recovery only by virtue of a fault in the discounting methodology used in its construction,[8] a problem that is easily corrected.

As shown here, inflation indexing in and of itself does not provide a recovery system superior to the current system. The severe nonneutrality of CCRS, both among assets and as between investment and consumption, cannot be corrected by indexing. And, despite the absence of indexing in ACRS, current law is more nearly neutral, both among assets and with respect to capital in the aggregate, than is CCRS at inflation rates at least as high as 10 percent. At more reasonable levels of inflation, around 5 percent, ACRS is almost the equivalent of expensing for many assets.

The NCRS, or a system constructed along these lines, holds the promise of achieving important goals of constructive tax reform. In its basic structure, it would make dramatic moves toward full cost recovery, without differentiating in this respect among assets with differing recovery periods. With its indexing provisions corrected to account for inflation in the first year in which deductions would be allowed, it would ensure achievement of these neutrality goals by preventing erosion of cost recovery allowances by inflation. As a somewhat more extended recovery system than ACRS, it would somewhat reduce the severity of the problems resulting from front-end loading of capital recovery. Because each year's deductions would be smaller in amount per dollar of capital outlay than the ACRS allowances, it would tend to increase the amount of each year's deductions that could be effectively used each year and reduce the amount that would have to be carried over to succeeding years. As an extension of ACRS, it would create far fewer transition problems than would a move toward a substantially different cost recovery system, such as CCRS. Moreover, it would generate additional tax revenue in the early years of its implementation, compared with ACRS-ITC, and thereby make possible desirable revenue-losing tax revisions which might otherwise have to be foregone under the constraints of a revenue-neutral reform program. One problem that would have to be dealt with in any system that lengthens recovery periods is the cash flow issue for investors. Notwithstanding the maintenance of present values in NCRS, it would reduce cash flow in the short run and would therefore have somewhat of a negative effect on those investors dependent on this cash flow.

There are problems, to be sure, in the design of NCRS. With extended recovery schedules, satisfying the condition that the present values of the allowances for all types of property are the same and just equal to the price of each property depends critically on whether the real after-tax rate of return used in the discount function is appropriate. If the recovery schedules are based on the assumption that the correct rate is 3.5 percent, as in the Kemp-Kasten proposal, the present value of the allowances will exceed 100 percent of the capital expenditure if the actual rate is less than 3.5 percent and fall short of it if the actual rate is greater than 3.5 percent. The extent of the shortfall or excess is not likely to be severe if the rate chosen to determine the write-off schedules is in the neighborhood of prevailing real after-tax rates of return. The statistical record suggests that with a 3.5 percent rate, this is likely to be the case, but it should be recognized that the results with NCRS are likely to diverge from the more nearly optimal ones that would be obtained with true expensing.

NCRS fails to address, but no more than ACRS-ITC or RCRS, the inadequacy of capital recovery systems that gear the beginning of cost recovery to the taxable period in which the property is placed in service rather than to the period in which the costs of acquiring it are first incurred. Although it would somewhat alleviate the problem of wasted deductions, NCRS would not ensure that the present value of deductions that must be carried forward would be the same as if they were fully effective in the year in which they are allowed. A similar problem would arise in the indexing provisions in the plan, which, as presently proposed, would not index deductions that have to be carried forward. These deficiencies could be readily overcome by appropriately increasing and indexing the amount of deductions that are carried forward, although this would, of course, increase the complexity of the system.

The NCRS proposal as it now stands leaves a number of important issues to be dealt with. Notwithstanding, it is an imaginative and innovative measure for moving the income tax in the direction of the neutrality that is needed to prevent the tax from distorting market signals and impairing economic efficiency. The administration would be well advised to replace the retrogressive CCRS with the highly promising NCRS as a major step toward true expensing.

A NOTE ON PRESENT VALUES

All present value calculations in this report follow a half-year convention whereby the first year flows are discounted by the appropriate discount rate to the one-half power, second-year flows to the one and one-half power, and so on. This is in contrast to the frequent practice of discounting the first-year flow by a full year, or to the Treasury practice of not discounting the first year at all. The logic of the latter is unintelligible, while the half-year convention is more accurate than full first-year discounting.

NOTES

1. For many years, much of the scholarly tax literature ignored the first-level nonneutrality, the income-tax bias against saving and investment in general, and concentrated on the second-level nonneutrality, the effect of differential provisions in the tax law in distorting the allocation of capital. This may be likened to focusing attention on preventing petty theft while doing nothing to prevent grand larceny.
2. ERTA did not aim solely at offsetting the erosion of capital recovery allowances by inflation. Its basic purpose was to reduce the inherent bias in the income tax against committing saving to investment in depreciable property.
3. The need for this qualification rests on the importance of the recovery pattern, as well as period.
4. For a demonstration that economic depreciation necessarily violates the conditions of tax neutrality, see Ture 1984:51–59, 72–76.
5. This loss in market value, equal to the reduction during the period in the present value of the remaining income the asset is expected to produce, depends on a myriad of factors which have differing effects from time to time, asset to asset, and taxpayer to taxpayer. The notion that any cost recovery system prescribed for the income tax will come anywhere near close to measuring economic depreciation boggles the mind.
6. Curiously, although CCRS has fewer classes than RCRS—one would want many times this amount if economic depreciation were the goal—as well as shorter recovery periods, notwithstanding the defense advanced for the original proposal, the administration proposal is accompanied by the verbatim incorporation of much of the Treasury text in favor of economic depreciation.
7. As presented in the administration's and Kemp-Kasten's proposal discussion of the FAST, the inflation indexing in both CCRS and NCRS is imperfect. Neither provides for indexing in the first year the asset is placed in service. This implicitly assumes that the inflation rate is always zero in that first year, only beginning in the

succeeding year. The consequence of this treatment, is that the present value of the allowances for any given class of property under RCRS or NCRS decreases as the inflation rate increases. If the first year's allowances were indexed for inflation, on the other hand, the present value of the allowances for any given class of property would not vary with the inflation rate.

8. The NCRS recovery schedules are designed with the use of a discount function which adds the inflation rate to the real after-tax rate of return to find the total rate of discount, that is, $(1 + r + z)$ where r is the real after-tax rate of return and z is the expected inflation rate. The correct inflation-adjusted discount rate, however, is the product of 1 plus the real after-tax inflation rate and 1 plus the inflation rate, i.e., $(1 + r) \times (1 + z)$.

REFERENCES

U.S. Treasury Department. 1984. *Tax Reform for Fairness, Simplicity, and Economic Growth.* vol 2. Washington, D.C.: U.S. Government Printing Office.

Ture, Norman B. 1984. "The Accelerated Cost Recovery System: An Evaluation of the 1981 and 1982 Cost Recovery Provisions." In C. Walker and M. Bloomfield, eds., *New Directions in Federal Tax Policy for the 1980s.* Cambridge, Mass: Ballinger.

10 BUDGETS, DEFICITS, TECHNOLOGY AND ECONOMIC GROWTH

Michael Boskin

The topic of this chapter is the potential impact of budget policy and deficits on technical change and economic growth. Writing at a time when the government budget in Great Britain was not only infinitesimally small relative to today's U.S. government budget but a much smaller share of their economy, British statesman William Gladstone remarked that budgets are not merely affairs of arithmetic but in a thousand ways go to the root of the prosperity of individuals, the relations of classes, and the strength of kingdoms. Despite the belief of some people that how nations finance their government's spending does not matter, my beliefs are somewhat closer to Gladstone's than to those of people who think that deficits matter not at all.

A basic point very dear to my heart, since I have been writing a book on it for two years, is that the federal government budget does not measure taxes or spending properly. There are a large number of items excused by law from inclusion in the federal government's budget. There are no adjustments from par to market values or for inflation. The federal government of the United States is one of the few governments of advanced economies that does not keep a separate capital account, so it treats spending on bridges the same as purchasing lettuce for the White House kitchen. I could go on and on. But lest you think that it is purely an historical accident, and I do believe that most of the accounting procedures are historical accidents, I commend to you the little heralded, but most insidious feature, of the 1983 Social Security Amendments which decree that in a decade Social Security will be removed from the budget of the U.S. government. Social Security is only the second largest revenue source and, depending on how you measure it, the second or first largest spending item. So what we include in the budget, the numbers we come to hear all the time, can be quite inaccurate.

Some other basic points are revealed in the following tables. Table 10−1 reveals that federal spending, as a share of gross national product, has risen substantially in recent years. Spending has gone up a lot, not down, under President Reagan. While there has been a defense build-up, the major culprits are interest on the debt, Social Security, and Medicare.

The single most important point to remember about the current level of fiscal activity and the way we are financing it, is that the large current and prospective deficits (we can argue about whether that is going to be $180 billion or $230 billion next year), roughly 5 percent of gross national product, and the overwhelming majority of American net private saving, are unprecedented in the United States in peacetime at close to full employment. The United States has had, and probably should have, such deficits during wars and at times of deep recession. What is unusual, and what is a matter of concern, is that we are moving into uncharted waters. Like most people, I have my own views of the way a prodeficit policy, a spending increase, or a tax cut, affects the economy, but extrapolating to this level over a long period of time may be somewhat dangerous.

The first way the federal government can affect technical change, research and development, and investment expenditures is via its direct spending decisions. It raises revenue by taxing or by borrowing, and it turns around and spends it. The first thing we might ask is what the federal government is really doing on the spending side. There are some conflicting opinions depending upon how one really tries to define investment, the conceptual issue of how one treats federal government payments to support state and local expenditures on capital projects which may have occurred anyway and the like. In the first Reagan administration there seemed to be a slight increase in federal government physical investment, heavily concentrated in military investment; and a slight decrease in federal R&D expenditures. And the major story in terms of R&D and investment was the substantial drop in

Table 10–1. Continued Growth of Federal Outlays

Federal Budget Outlays, 1966–1984

Fiscal Year	Current Dollars (Billions)	Outlays as a Percentage of GNP
Ave. 1966–1970	169.9	20.2
Ave. 1971–1975	255.7	20.4
Ave. 1976–1980	456.2	21.7
1981	657.2	22.8
1982	728.4	23.8
1983	796.0	24.7
1984	851.8	23.8
1985 Est.	959.1	24.8

Source: The Budget of the United States, Fiscal Year 1986.

quasihuman capital education and training expenditures. We cannot properly measure output by the cost of inputs. I do not know if the increased military hardware purchases and R&D in the military or the decrease in some of the training and manpower and education expenditures are worthwhile on a cost/benefit basis. We should not just look at expenditures and assume returns and assume thus that a decline from any particular previous level is appropriate or an increase is desirable and yields a rate of return commensurate to what could have been gotten in the private sector.

There has been an enormous amount of concern and alarm and it is rather clear that the United States has launched on a major military build-up in terms of physical investment and there has been a slight decrease in federal investment overall, heavily concentrated in grants for capital expenditures for state and local governments. These features are revealed in Table 10–2. A substantial amount of acceptable econometric research suggests that the investment at the state and local level overwhelmingly would have gone on in the absence of the federal government expenditures and that in net what has really happened is that Americans have substituted federal government spending for state and local financing of most of that physical investment.

Beyond the direct impact in terms of how the government spends its resources, there is obviously the potential impact of how it raises its

Table 10–2. Federal Outlays by Category, Selected Years ($ Billions of Constant 1972 Dollars; Percentages of GNP in Parentheses).

Fiscal	Total	Physical Investment			R&D		Payments to	Net	All
Year	Outlays	Defense	Nondefense	Grants-in-Aid	Defense	Nondefense	Individuals	Interest	Others
1949	85.3	4.0	1.8	0.9	2.0	0.5	17.6	8.4	50.4
	(17.3)	(0.8)	(0.4)	(0.2)	(0.4)	(0.1)	(3.6)	(1.7)	(10.2)
1958	135.3	25.2	1.9	2.8	4.7	1.4	30.1	8.4	60.8
	(19.9)	(3.7)	(0.3)	(0.5)	(0.7)	(0.2)	(4.4)	(1.2)	(8.9)
1968	224.6	30.2	3.4	7.6	11.4	10.2	60.1	13.5	88.1
	(21.2)	(2.9)	(0.3)	(0.7)	(1.1)	(1.0)	(5.7)	(1.3)	(8.3)
1978	300.4	15.8	4.4	11.0	7.6	7.9	144.7	23.6	85.4
	(20.9)	(1.1)	(0.3)	(0.8)	(0.5)	(0.5)	(10.0)	(1.6)	(5.9)
1984	363.5	31.3	3.2	9.0	11.4	5.9	182.0	44.6	76.1
	(22.2)	(1.9)	(0.2)	(0.5)	(0.7)	(0.4)	(11.1)	(2.7)	(4.6)

Sources: Office of Management and Budget, Budget of the United States, selected years, and appendixes.

resources. If the federal government in addition to the amount it withdraws from the private sector does so in a way that comes heavily at the expense of research and development or investment, then it may impinge upon technical change and economic growth in the long run.

There are two really important potential ties between the rate of investment and the rate of technical change and technical progress which underlies our long-term rate of economic growth per person about which we really have insufficient information. The first, the so-called embodiment hypothesis, is an old subject that Dale Jorgenson researched (see Jorgenson 1966). It conjectures that new technology is so costly to embody in old capital that it either absolutely or disproportionally must be embodied only in new capital and therefore the more rapid the rate of increase of the capital stock, the higher our rate of investment, the more rapidly new technology will be disseminated throughout the economy and conversely the greater the derived demand for the generation of that new technology.

Second, a hypothesis associated with my colleague, Kenneth Arrow (1962), is that in the process of investment, and also obviously in the process of research and development, new products and new processes become available which otherwise would not have become available. And so in the act of investing we learn by doing and therefore for that reason, among others, the social return to investment is much higher than the private return and investment potentially feeds back positively on the rate of technological progress. If either of those conjectures is correct and quantitatively important then not only should we be paying attention to the direct effect on however we want to try to measure technical progress, it would be important to be able to quantify the impact of the budget, taxes, and deficits on physical investment. Again, those links which are conceptually vital to the tie between high investment and a high growth economy at a given level of aggregate demand will require much further research.

Table 10–2 reports historical data on the budget of the United States which is an attempt by the Office of Management and Budget to develop a separate measure of major physical capital investments, R&D expenditures, payments for individuals, interest on the debt, and all other outlays. For example, grants in aid for physical capital investment to state and local government. The major thing to note is that expenditures for the conduct of research and development in real dollars, let alone as a share of GNP, are substantially less than that in the 1960s and that the overwhelming bulk of that decline is in the nondefense area in real spending on research and development.

My concern is that the government must be involved in the support of basic research because of the difficulty of private investors appropriating all of the returns. Whether we are doing enough R&D at the federal level and financing enough of it is an issue I will not get into. Obviously there is a substantial case for some government involvement at least in the financing of R&D because of the potential public good nature of research and development that would suggest that private firms cannot appropriate all of the returns and would therefore underinvest in R&D.

The spending control gap.

It is constructive to go back and look at 1981 at what we were spending as of 1981 when President Reagan took office and what he proposed in his March 1981 program as a target for 1985. Table 10–3 reports these

Table 10−3. The Spending Control Gap: Reagan Administration's Original Plan for Fiscal 1985 versus Current Projection for Fiscal 1985 (Outlays as Percentage of GNP).

	Actual 1981	March 1981 Target for Fiscal 1985	Current Fiscal 1985 Projection	Difference
National defense	5.5%	6.5%	6.5%	
Benefit payments				
Non-means-tested	11.4	8.3	9.5	+1.2
Means tested		1.7	2.1	+0.4
Grants	3.0	1.0	1.5	+0.5
Other	0.6	1.2	1.5	+0.3
Net interest	2.3	1.5	3.3	+1.8
Undistributed cuts		−1.0		+1.0
Total	22.8%	19.2%	24.4%	+5.2

as shares of GNP by category and the difference from actual levels. And we see that the major change is that we have had about a one percentage point build-up in defense spending, a very substantial increase in non-means-tested spending programs, and that means-tested spending has declined as a share of GNP, but not as far as the president had originally proposed. The federal government has sharply reduced its grants in aid to state and local governments, probably not unreasonably given the surplus that state and local governments are running. But I will remind you of another of the conceptual difficulties of budget reporting: a nontrivial share of that state and local surplus is in their retirement funds which are simultaneously accruing liabilities. Net interest has increased substantially rather than gone down and many of you may not remember, but if you go back to 1981 there actually was 1 percent more of undistributed cuts which had not been figured out yet. So the administration was trying to cut spending to 19 percent of GNP. It is running 24 percent and therefore we have this 5 percent gap.

If you look at Table 10−4 you will see the CBO's baseline budget projections assuming no changes in the laws governing taxes or entitlement spending and including the off-budget of spending $15 billion (just to give you an idea of the problems that we have in government budgeting, we have an official category we now call off-budget spending, though it is only a minute fraction of the total off-budget spend-

Table 10–4. CBO Budget Projections-Baseline (by Fiscal Year).[a]

	1983 Actual	1984 Actual	1985 Base	1986	1987	1988	1989
Total deficit ($ Billions)[b]	208	185	210	212	229	243	264
Deficit as percentage of GNP	6.4	5.2	5.5	5.1	5.1	5.1	5.1
Debt held by public as percentage of GNP	35.4	36.7	39.6	41.9	43.9	45.8	47.5

a. Assumes no change in laws governing taxes or entitlement spending.

b. Includes off-budget deficit of $15 billion or 0.3 percent of GNP per year.

ing). Also at the bottom of that table is the debt held by the public as a percentage of the GNP and where it is expected to go under current law. Clearly these are very large deficits. The administration forecasts much lower deficits partly because it is somewhat more optimistic in what it was going to be able to do on the spending side, but heavily because it assumes that interest rates will fall substantially over the next few years (down to 5 percent). If you took the administration's forecast for the 1989 or 1990 budget and you assumed that interest rates were going to be what they are today, you would add about $70 billion to that fiscal year's deficit.

Table 10–5 points out that if you think the current fiscal dilemma faced by the United States will be resolved in the course of this decade and then we can breathe a deep sigh of relief, we have impending a major conundrum with respect to the financing of our Social Security and Medicare systems. We had this tremendous, bipartisan Social Security reform in 1983. Column 1 shows pre-1983 amendments and the Social Security actuaries' intermediate assumptions. They make assumptions about economic growth, inflation, unemployment, life expectancy, fertility, how many workers there will be relative to the number of retirees, and the like, and from these assumptions they calculate the likely future condition of the retirement fund, disability fund, and hospital insurance fund (which is also part of Social Security). Prior to 1983 the Old Age and Survivor Disability Fund (OASDI),

Table 10–5. The Long-Term Social Security and Medicare Deficit.[a]

	Pre-1983 Amendments	1983 Amendments Intermediate Assumptions	Indexing of Taxable Amounts or Dissipation of Surplus	Indexing and Dissipation of Surplus
OASDI	−1.80	0.02	−0.60	−1.20
HI	−5.21	−5.21	−5.21	−5.21
Total	−7.01	−5.19	−5.81	−6.41

a. Effect of 1983 amendments on actuarial balance of OASDI and hospital insurance fund—alternative assumptions, seventy-five year period (percentage of taxable payroll).

Source: Author's calculations plus 1983 Annual Report of the Trustees of the Social Security Administration.

which is commonly called Social Security, projected an average 1.8 percent deficit over the seventy-five-year projection period as a fraction of taxable payroll. However, this average masks a huge surplus in retirement and disability starting about in 1990 that would run for about twenty-five years, then start to decline. We would have absolutely enormous deficits when the baby boomers retire early in the next century. So that 1.8 average masks deficits of 5 to 8 percent in 2030 and surpluses that are projected to build up to 7 years' payments in the retirement and disability funds by 2015, an amount equal to the current (privately-held) national debt. Prior to the 1983 amendments (the Greenspan Commission was told to leave it to someone else), it was not commonly pointed out that the long-term deficit in the hospital insurance fund was several times that in retirement and disability. That is primarily because the tax rate in the hospital insurance fund was very low and not scheduled to rise as the tax rate was for retirement and also because of the presumed increase in the cost of medical care and changing demographics.

This 1.8 percent disappears under the intermediate assumptions. But the 1983 amendments did two things that are going to be incredibly difficult to live by politically. One is that now we are taxing Social Security benefits for well-off retirees. Half of Social Security benefits are included in taxable income at an income above $32,000 a year. But that number is not indexed. So eventually the United States will be taxing middle-class retirees and it is unclear that it will be possible to keep up that unindexed portion or keep up with not indexing the exempt amounts. Indexing the exempt amount will mean losing about a third of the net financial solvency gained in the OASDI system with the 1983 amendments.

The second thing is the Social Security system has never been able to run a surplus. As soon as the system starts running substantial surpluses, the government has either raised the benefits or, less commonly, lowered tax rates. And it is going to be very interesting if these funds are kept separate and the retirement fund is not allowed to bail out Medicare when it starts running very large deficits in the next decade if the funds are kept separate as was originally intended and this surplus is used to try and dissipate the need for a huge tax increase early in the next century. Doing that will result in a surplus in OASDI five to seven years' worth of Social Security benefits and such a large surplus that it will wipe out the national debt. That is so large a Social Security surplus that the Social Security Administration would own all the government bonds. Running even a higher surplus or getting some spending under control would raise the question of whether we Americans want the Social Security system to start owning other assets, such as corporate shares. But have no fear. The hospital insurance fund will bail us out of that conundrum. We face enormous long-term deficits in Medicare which will test not only our political acumen but the moral fiber of our country in the next twenty years.

IMPACTS OF BUDGETS AND DEFICITS

One thing that all economists would no doubt agree upon is that *ultimately* at full employment, large deficits, net of the interest component, the so-called primary deficit, run continuously relative to the size of the economy, for a very long period *must be inflationary.* There is no need for the current budget deficit to be inflationary for the next year or two. But if we run primary deficits of several percent of GNP for a very long period of time, some strange things are going to happen. And to analyze that, examine Table 10−6 and the following equations, which indicate where the ratios of privately held national debt, what is called debt held by the public or by the private sector as opposed to the Federal Reserve and government agencies, is headed if we continue a deficit as a share of GNP net of the interest component and run the real interest rates and growth rates that are being forecast.

Let D represent the debt/GNP ratio, d the deficit (net of interest)/GNP ratio, r the real interest rate, g growth of real GNP. Then, by definition

$$\dot{D}_t = d_t + (r_t - g_t)\,dt$$

for a fiscal program with constant d, and constant r and g, D will evolve toward en equilibrium D_e (if $g > r$) of

$$D_e = \frac{d}{g - r}.$$

The ratio of the federal government debt to GNP evolves through time depending upon this primary deficit and the relation between the real rate of interest and the growth rate. For example, if we start out with a positive national debt, and the real interest paid on the national debt exceeds the growth rate, then the interest payments will grow more rapidly than the GNP, and if nothing else has changed, eventually the interest payments will gobble up all of the budget, then all of GNP, then all of national wealth in an explosive pattern. In the more usual case of the growth rate exceeding the rate of interest, the ratio of debt to GNP will evolve according to the equation above.

Table 10–6 presents some estimates of two recent major fiscal episodes in the United States. First, we see the substantial increase in the equilibrium debt/GNP ratio toward which we were headed if fiscal policy had not been changed, in the 1975–1979 period, a period generally regarded as the beginning of the increase of the ratio of debt to GNP after the substantial postwar decline in this ratio. The second, and more important for our purposes, is where we are headed starting currently. We can see that under current projections, the ratio of debt to GNP is heading toward an equilibrium which is many times, not only current GNP, but the ratio of the entire value of the capital stock of the United States to GNP. This latter number is around 3, so it is clear that either the private sector will have to increase its wealth to income ratio by an enormous increase in saving, or the rest of the world will have to buy up the Treasury bills that the United States floats, or if neither of these alternatives is available

Table 10–6. Some U.S. Fiscal Episodes.

	1975–1979[a]	1984–1989	
D_o	23.4%	37%	
d, average	3.7	1.8	1.4
g, average	3.5	3.8	4.0
i (net of monetization)	6.0		
GNP deflator	7.2		
r	−1.2	3.6	2.4
$g - r$	4.7	0.2	1.6
D_e	79%	900%	88%

Note: D_t declined steadily from World War II to 1974.

and we persist in our current fiscal policies for the indefinite future, the Federal Reserve will have to buy up the bonds as the lender of last resort, thereby reigniting inflation. Can we Americans reasonably expect foreigners to continue to finance our deficits ad infinitum? It would be imprudent of us to operate on the assumption that this was possible, let alone desirable. Eventually, foreign firms and individuals will have a progressively higher fraction of their wealth in dollar-denominated assets, which will mean that further increases in dollar-denominated assets will be even riskier. Thus, we can expect the flow of foreign capital into the United States, ceteris paribus, to slow down. Nor do I foresee such a huge increase in our saving rate as to increase the capital output ratio by such a large amount. In short, the current fiscal policy, if continued, is either inflationary or unsustainable. This does not mean that we have no time whatsoever to deal with the problem, but the longer we delay in cutting spending or adopting a last-resort tax increase, the worse the problem will become as the interest burden rises still further.

Thus, my conclusion is that while we had an investment boomlet in the United States in 1984, and this in turn was largely the result of proinvestment tax policies enacted in the 1981 tax laws, eventually these large deficits will crowd out our own investment. Currently, they are probably playing some role in crowding out our net exports.

Table 10–7 indicates the decline in effective tax rates on marginal investment due to ERTA and TEFRA. This decrease was substantial and must be part of the explanation for the resurgence, once out of the

Table 10—7. Effective Tax Rates for All Assets, 1965—1982, Selected Years.

	Auerbach[a]	Hulten—Robertson[b]
1965	35.7	26.3
1970	49.7	52.3
1975	37.0	32.1
1980	31.9	33.1
1981	17.7	4.7
1982	24.6	15.8

a. Auerbach assumes a 4 percent real after-tax rate of return and forecasts inflation based on past values.

b. These estimates assume a 4 percent real after-tax rate of return and, for 1981 and 1982, a 6 percent rate of inflation.

Sources: Auerbach, Alan. 1983. "Corporate Taxation in the United States," *Brookings Papers on Economic Activity* 2.

Hulten, C. and J. Robertson. 1982. "Corporate Tax Policy and Economic Growth." Urban Institute Discussion Paper, December.

recession, of investment. Another reason for the resurgence is that the extended investment tax credit and the Accelerated Cost Recovery System are available only on domestic investment. This, in turn, has probably also been a cause of the decrease in U.S. capital exports, which again, has been part of the reason for taking some of the potential pressure of deficits off of interest rates. Table 10—8 reports our current gross saving, investment, and net saving and investment figures for the private economy. Clearly, while the United States has had an investment boomlet, our net national saving and net national investment are still quite modest relative to our history and to those of our major trading partners.

Finally, Table 10—9 reports alternative ways of viewing the share of society's resources devoted to the present as opposed to the future. It becomes clear that for a variety of reasons, the U.S. economy has been consuming a huge fraction of its net national product in recent years, and that our national saving rate has plummetted.

What can be concluded from all of these figures? My tentative conclusion is that we still have an investment problem in the United States, that we will probably have to finance investment by our own domestic saving, and that we have an equally severe saving problem.

Table 10–8. The Role of the Federal Deficit in National Saving and Investment (as Percentage of GNP).

	(1) Net Private Saving	(2) State-Local Surplus/ Deficit	(3) Federal Surplus/ Deficit[a]	(4) Net National Saving Saving	(5) Capital Outflow (–) or Foreign Inflow (+)	(4+5) Net Domestic Investment
1950–1959	7.2	–0.2	0.0	7.0	–0.2	7.0
1960–1969	7.9	0.0	–0.3	7.6	–0.5	7.1
1970–1979	7.1	1.0	–1.8	6.3	0.0	6.4
1982	5.3	1.0	–4.8	1.5	0.3	1.8
1983	5.9	1.3	–5.4	1.8	1.0	2.8
1984	7.4	1.4	–4.8	4.0	2.6	6.5

a. NIPA basis.

Table 10–9. Alternative Measures of National Saving Behavior, Selected Years.

Period	(1) National Income Accounts Basis Net National Saving (%)	(2) Household Consumption Rate (%)	(3) Government Consumption Rate (%)	(4) (2)+(3) (%)	(5) Household Consumption Rate out of NNP–Government Consumption (%)
1951–1960	6.9	67.6	17.0	84.6	81.6
1961–1970	7.5	65.4	20.4	85.8	82.0
1971–1980	6.1	67.1	21.0	88.1	84.9
1981–1984	3.3	68.4	24.4	92.8	90.5

Note: NIPA data adjusted to treat durable purchases (household and government) as saving and imputed rental flow on household durables plus government tangible assets on consumption. NNP adjusted to include imputed rent and exclude depreciation on household durables and government tangible assets.

Economic policies designed to encourage consumption at the expense of saving and investment have marked U.S. fiscal history far too long. It is time we consolidated those policies begun in 1954, continued in

1962, 1978, and 1981 in moving our tax policy toward neutral consumption taxation and our overall fiscal policy toward appropriately measured budget balance over periods longer than cycles in economic fluctuations to renew the prospect of a high saving and high investment economy which may well feed back on a higher rate of technological change and economic growth.

Let me repeat that many of the links in this chain of argument are based on less than firm footing. Much more research needs to be done, for example, on the determinants of saving and investment in the United States, on the potential links between investment and technical change. The United States badly needs a rededication of national economic policies toward promoting the country's long-term growth and international competitiveness, and these in turn will ultimately require much higher rates of national saving and investment in plant and equipment as well as research and development in the United States.

REFERENCES

Arrow, K.J. 1962. "The Economic Implications of Learning by Doing," *Review of Economic Studies*

Auerbach, Alan J. 1983. "Corporate Taxation in the United States," *Brookings Papers on Economic Activity* 2:451–513.

Hulten, C. and J. Robertson 1984. "The Taxation of High Technology Industries," *National Tax Journal* 37 (September).

Jorgenson, D.W. 1966. "The Embodiment Hypothesis," *Journal of Political Economy* (February).

11 FINANCING AMERICAN INVESTMENT IN NEW TECHNOLOGY

Benjamin M. Friedman

The current combination of fiscal and monetary policies in the United States is seriously hindering the U.S. economy's ability to take advantage of newly developing technology. Large federal government budget deficits, on a scale unprecedented in peacetime experience, are now absorbing more than one-half of all saving done by individuals and businesses in the United States. As a result, the volume of plant and equipment investment now being undertaken by the economy's business sector is well below what would normally occur in the advanced stages of a vigorous expansion. And because these large deficits are causing not only high real interest rates but also a high real dollar exchange rate, many of the nation's industries are no longer able to compete for sales abroad or in internationally exposed markets at home.

Under current tax and spending policies, these budget deficits will not shrink in the foreseeable future, even as the expansion eliminates the economy's remaining gap of underemployed labor and capital resources. A sufficiently expansionary monetary policy could soften the deficit's negative implications for the economy's capital formation, of course—at least for a year or two. After that time, however, this policy would amount to simply inflating away the government's debt. After the enormous cost the United States has sustained during the last five years to reduce the rate of price inflation, returning again to a high-inflation economy would hardly be sound policy. The right way to deal with an excessively expansionary fiscal policy is to change fiscal policy, not to adopt an inflationary monetary policy.

At the same time, counting on a combination of economic growth and cuts in a variety of small spending programs to narrow the deficit substantially is betting on an illusion. Under today's circumstances major deficit reduction can come only from (1) cuts in large entitlement programs like Social Security or Medicare, (2) a slower defense buildup, or (3) a broad-based tax increase. Seeking to address the problem without resort to any of these three measures is a recipe for continuing large deficits, continuing high real interest rates and real exchange rates, continuing weakness in business capital formation, and continuing failure in international economic competition. The best approach to solving these problems would combine all three measures, thereby minimizing the potentially negative side effects likely to be associated with any one.

Regardless of the specific elements of whatever new fiscal policy replaces the current one, it is now clear that some major change in fiscal policy is necessary if the U.S. economy is to take advantage of newly developing technology. Without a new direction in fiscal policy, the pace of productivity growth and product change in the United States will increasingly lag behind not only foreign competition but also traditional American aspirations for economic well-being.

THE DEFICIT PROBLEM

With the current economic expansion in the United States now beginning its third year, the persistence of federal government budget deficits on an unprecedented scale has clearly become the chief distinguishing characteristic of this business cycle—both economically and politically. The budget submitted to Congress by President Reagan in February 1984 acknowledges that, after a small improvement in 1984, the fiscal 1985 deficit will exceed even the record $208 billion mark posted in the immediate postrecession environment of 1983. More important, the president's budget indicates that even with a strong economy, and even after $50 billion of proposed spending cuts, the 1986 deficit will show no improvement over 1984.

The basic fiscal policy experiment run in the United States during 1983 and 1984 has been to determine whether the combination of vigorous economic growth and selected cuts in nondefense spending programs would be sufficient to eliminate the government's budget deficit, or even to shrink it significantly. The answer is no. Instead, the result has been the emergence of ever larger deficits, which as the economy expands are increasingly deficits at full employment. The Reagan administration's most recent estimates project that, in the absence of fiscal policy changes, the deficit will remain at about today's level through the remainder of the decade. By contrast, the Congressional Budget Office, which has assumed that real economic growth during 1986–1990 will average 3¼ percent per annum instead of the 4 per cent per annum assumed by the administration, projects that without policy changes the deficit will grow to nearly $300 billion by the end of the decade.

Sustained government deficits of this magnitude, either in dollars or in relation to gross national product, will be unprecedented in U.S. peacetime experience. Table 11–1 shows the record of the federal government's budget deficit since 1950, both in dollars and in percentages of GNP. Despite the often expressed claim that the government's budget has "always" shown a large deficit, in fact persistent deficits larger than ½ percent of GNP have been a feature of U.S. fiscal policy only since the 1970s. Until 1982 the deficit had exceeded 3 percent of GNP only during 1975 and 1976, in the wake of the severe 1973–1975 business recession. Analogous effects of the 1981–82 recession pushed the deficit to more than 4 percent of GNP in 1982, and more than 6 percent in 1983.

Unlike few relatively isolated episodes of large deficits in the past, which largely reflected the shortfall of tax revenues and increase in transfer payments due to declining employment, incomes and profits in times of recession, the deficits now projected for the balance of the 1980s will increasingly represent a budget that will be unbalanced even at full employment. Table 11–2 shows the federal deficit record since 1970, adjusted for just these cyclical effects to leave only the deficit that the government would have run if the economy had operated at 6 percent unemployment.

The difference between the actual and high employment budget concepts, shown in Tables 11–1 and 11–2, respectively, is especially important in determining what magnitudes constitute the outer limits of the U.S. economy's prior experience. In 1975–76, for example, the actually combined deficit of $126 billion corresponded to a high

Table 11–1. U.S. Government Budget Deficits, 1951–1985.

	Deficits in Billions of Current Dollars	Deficits as Percentages of GNP
Average, 1951–1960	$ 1	0.0%
Average, 1961–1970	6	0.5
Average 1971–1980	31	2.4
1971	23	2.2
1972	23	2.1
1973	15	1.2
1974	6	0.4
1975	53	3.6
1976	73	4.5
1977	54	2.9
1978	59	2.8
1979	40	1.7
1980	74	2.9
1981	78	2.8
1982	128	4.2
1983	208	6.4
1984	185	5.2
1985 Est.	222	5.7

Source: Office of Management and Budget.
Notes: Deficit totals include "off-budget" outlays. Years indicated are fiscal years.

employment deficit of only $57 billion. Similarly, in 1981–82 the combined actual deficit of $206 billion would have been only $70 billion at high employment. In comparison to the economy's size, the largest high employment deficit run during the 1970s was 2.1% of GNP in 1976. Beginning in 1983, every new year is now marking a new record on this basis.

What is extraordinary about the U.S. government deficits now projected for the remainder of the 1980s, therefore, is not just that they will be large but, more important, that they will represent a fundamental imbalance between the government's revenues and its expenditures. It is not possible to dismiss them simply by assuming that rapid growth will quickly restore the economy to full employment. The projected

Table 11–2. Deficits Adjusted to 6 Percent Unemployment, 1971–1985.

	Deficits in Billions of Current Dollars	Deficits as Percentages of GNP at 6 Percent Unemployment
Average, 1971–1980	$ 27	1.8%
1971	20	1.9
1972	22	1.9
1973	24	1.9
1974	13	1.0
1975	23	1.5
1976	34	2.1
1977	35	1.9
1978	39	1.9
1979	21	.9
1980	36	1.4
1981	24	.8
1982	46	1.4
1983	91	2.6
1984	118	3.2
1985 Est.	155	3.8

Source: U.S. Department of Commerce, Office of Management and Budget.
Notes: Deficits are on a unified budget basis. Years indicated are fiscal years.

deficits are increasing by deficits at full employment, and in the absence of a return to full employment the deficits that actually emerge will only be larger. The issue now facing U.S. fiscal policy is not the familiar one of the role of automatic stabilizers, or even the desirability (or lack thereof) of temporary active deficits as discretionary stabilizers, but rather the effects of sustained deficits at full employment as a permanent feature of the economy's ongoing development. Among the most important of those effects is the impediment that such deficits will place in the way of the capital formation that would otherwise provide the vehicle for bringing newly developing technology into the U.S. economy's production systems.

DEFICITS AND CAPITAL FORMATION

The question now confronting the U.S. economy is what influence these enormous high employment deficits will have on the financial markets and hence on nonfinancial economic activity. The most likely answer is that, increasingly during this period, the pressure created by financing the government deficit will squeeze both private financing and private investment. As a result, although prospects are now good for continuing business growth (at least in the near term), the United States will be laying a poor foundation for economic development during the remainder of the 1980s and beyond.

At the most fundamental level, what is wrong with today's large government deficit is that it is swallowing more than one-half of all net saving done by the economy's private sector. The economy's net private saving, consisting of personal saving plus corporate retained earnings, represents the share of total output that the private sector as a whole makes available to finance new investment beyond what is necessary simply to replace depreciating stocks of business and residential capital. Despite substantial variation since World War II in such factors as tax rates, price inflation, real rates of return, and income growth trends—all of which could in principle affect saving behavior—the U.S. economy's net private saving rate has remained very steady throughout this period. Its postwar mean has been 7.2 percent, with a standard deviation around the mean of only 1 percent, and it has displayed no significant time trend during this period (once the data are corrected for cyclical variation). The saving rate has varied in a modestly procyclical pattern, however, which accounts for the slightly higher than average level during the 1960s and (in part) for the distinctly lower than average level thus far during the 1980s.

If government budgets were always balanced (and if the foreign account were balanced too), the share of the economy's output available for net capital formation would simply be the share set aside as net private saving. Given the experience since World War II, that would mean a relatively steady 7 percent of GNP over time. In the presence of government surpluses or deficits, however, what is available for net investment is net private saving plus any government surplus, or less any government deficit.

In recent years public sector saving and dissaving has played an increasingly prominent role in shaping the U.S. economy's overall saving and investment balance. Table 11—3 shows the U.S. balance of net saving and net investment since 1951, in percentages of gross national

Table 11–3. U.S. Net Saving and Investment, 1951–1984.

	1951–1960	1961–1970	1971–1980	1981	1982	1983	1984
Total net saving	6.9%	7.5%	6.1%	5.2%	1.6%	1.8%	4.0%
Net private saving	7.2	8.0	7.1	6.1	5.4	5.9	7.4
Personal saving	4.7	4.7	4.9	4.6	4.4	3.6	4.3
Corporate saving	2.5	3.3	2.2	1.4	1.0	2.3	3.2
State-local govt. surplus	−0.2	0.1	0.9	1.3	1.1	1.3	1.4
Federal govt. surplus	−0.2	−0.5	−1.9	−2.2	−4.8	−5.4	−4.8
Total net investment	7.0%	7.5%	6.3%	5.4%	1.6%	1.8%	3.8%
Net foreign investment	0.3	0.5	0.1	0.2	−0.2	−1.0	−2.6
Private domestic investment	6.7	7.0	6.2	5.2	1.8	2.9	6.4
Plant and equipment	2.7	3.5	3.0	3.1	2.0	1.5	2.4[e]
Residential construction	3.2	2.5	2.5	1.3	0.6	1.8	2.4[e]
Inventory accumulation	0.8	1.1	0.7	0.9	−0.9	−0.4	1.6
Memoranda: Capital consumption	8.9%	8.5%	9.9%	11.2%	11.7%	11.4%	11.0%
Gross private saving	16.1	16.4	17.0	17.2	17.1	17.3	18.4

Source: U.S. Department of Commerce, and author's estimate.

Notes: Data are averages (except for 1981–1984) of annual flows, as percentages of GNP. Total net saving and total net investment differ by statistical discrepancy. Detail may not add to totals because of rounding.

product. Since the 1970s state and local governments, in the aggregate, have run ever larger budget surpluses on a consolidated basis, as current pension surpluses have grown faster than operating deficits. By contrast, during this period the budget deficits run by the federal government have grown progressively larger in relation to GNP. These two trends have been in part offsetting, but increasingly unequal. By the early 1980s the federal government's deficit had grown far beyond the aggregate surplus of state and local governments. Under any currently relevant projections, it will remain so.

The U.S. economy's *total* net saving, consisting of the relatively steady net private saving plus government saving or dissaving, has therefore declined sharply since the low-deficit days of the 1950s and 1960s. As Table 11–3 shows, the economy's total net investment (which differs from total net saving only by a fairly small statistical discrepancy) has, of course, declined in equal measure. Because of a change from positive net foreign investment on average before the mid-1970s to negative net foreign investment on average thereafter, the deterioration of net domestic investment has been less severe than that of total net investment. Even so, net domestic investment has declined from 7.0 percent of GNP on average during the 1960s to 6.2 percent on average during the 1970s, and only 4.1 percent thus far during the 1980s. All components of net domestic investment—business plant and equipment, residential construction, and business inventory accumulation—have shared in this decline. The absorption by government of so much of the economy's private saving is the principal reason why fixed investment in the United States is now relatively weak, even under the impetus of a vigorous business expansion.

Despite the widespread publicity received by the 1984 gains in business capital spending from the immediate postrecession depths of the prior year, in any broader historical context the U.S. economy's investment rate today is not high but low. Business capital spending has, of course, recovered sharply from the depths it reached at the bottom of the 1981–82 recession. Gross spending for new plant and equipment in 1984 was $426 billion, up 21 percent from the $353 billion that business spent the year before. More important, *net* investment—that is, investment spending over and above what was necessary to replace worn-out older structures and equipment—rose even more dramatically, from $50 billion in 1983 to an estimated $90 billion in 1984. What is missing in this familiar analysis, however, is any sense of how high is up. The development that has lately attracted so much attention is the large increase in investment spending from the recession-depressed level of a year or two ago, so that to place this latest recovery in perspective it is necessary to recall just how depressed that level was.

As Table 11—3 shows, the $50 billion that American business spent on net additions to its productive capacity in 1983 amounted to only 1.5 percent of that year's GNP. That performance made 1983 the second-worst investment year in the nearly four decades since World War II. Only in 1975, after the sudden fourfold increase in oil prices had not only squeezed economic activity but also rendered obsolete many kinds of energy-intensive capital goods, was the nation's net investment rate as low as 1.5 percent. The much heralded improvement to an estimated 2.4 percent of GNP last year was welcome, to be sure, but climbing off the basement floor is not the same as scaling new heights.

The most visible reason for the continuing weakness of business investment in the United States is the enormous drain on the economy's saving represented by the persistence of massive—and growing—government deficits on a full-employment basis. Indeed, were it not for the huge inflow of investment capital from abroad, the U.S. economy's combined rate of investment in business and residential capital would be even smaller. In 1984 this net inflow of funds from abroad was $95 billion, about enough to finance one-half of the government's deficit. The drawback of relying on this source of financing, however, is that the economy's net foreign capital inflow is just the mirror image of its deficit on current account in the balance of international payments. The darker side of last year's record inflow was an astonishing erosion in the ability of American businesses to compete for sales either abroad or against foreign competitors at home, primarily as the consequence of the dollar's sharply higher international exchange value, which resulted in a record $123 billion trade deficit. While this foreign capital inflow did partly rescue the economy's investment sector, therefore, it did so only by decimating the economy's internationally-competitive sector, and at the same time driving the United States into net debtor status.

In the context of this historical experience of the U.S. economy's balance of saving and investment, the implications of the nation's current fiscal policy are clear enough. The Reagan administration has estimated that its $222 billion federal deficit estimate for fiscal year 1985 will represent 5.7 percent of the year's GNP. Even after allowing for the $40 billion to $50 billion surplus that state and local government units collectively will probably run, therefore, the government sector as a whole will still siphon away more than one-half of the private sector's net saving. Prospects for subsequent years, throughout the remainder of the 1980s, are little different under current policies. In the absence of major policy changes, real interest rates will remain high, and only the economy's faltering international-trade sector will enable net business capital formation even to approach postwar historical norms, much less surpass them.

SOME EASY BUT WRONG ANSWERS

When problems are as serious as this analysis indicates that the U.S. government deficit problem will soon be, there are rarely any easy answers. If there were, they would already have been adopted and the problem would not be so serious after all. Nevertheless, hard problems that are resolvable only by hard policy choices almost inevitably elicit suggestions that no such tough choices are necessary, either because some easy solution is readily available or because the ordinary course of events will provide one on its own.

The threat to American business capital formation posed by the U.S. government deficit in the 1980s is no exception in this regard. Three supposedly easy answers have dominated much of the public policy discussion of this issue to date. All are unsatisfactory.

Easy answer 1 is that the lower tax rates legislated in the Economic Recovery Tax Act of 1981 will sufficiently stimulate personal saving to enable the economy to finance both enlarged government deficits and an increase in net capital formation.

Such an outcome is conceivable, of course, but it is unlikely for several reasons. To begin, a priori reasoning alone cannot even indicate the *direction,* not to mention the magnitude, of the effect of higher after-tax returns on personal saving. The reason for this seemingly startling result is that higher after-tax yields affect the incentive to save in two potentially offsetting ways. Higher yields increase the reward, in terms of spending made possible later, for not spending today. This change in the terms on which consumers may substitute future for current spending increases the relative attraction of future spending, and hence unambiguously *encourages* saving out of current income. Higher yields also increase the rate at which savings grow, however, thereby making it possible to maintain desired spending levels in the future on the basis of less put aside today. This increase in lifetime total spending possibilities enables consumers to spend more both currently and in the future, and hence unambiguously *discourages* saving out of current income. Which of these two opposing effects is predominant can be determined only by resort to empirical evidence.

The problem here, however, is that the evidence on this question is ambiguous to say the least. Many important economic relationships have theoretically undetermined directions as well as magnitudes, but in the case of the effect of after-tax yields on saving the available evidence still provides essentially no ground for confidence about even the sign, much less the magnitude, of the net effect. Some studies have shown positive effects of a magnitude that would be meaningful in the context of the prospective deficit problem, while many others have shown no noticeable effect at all. At best, relying on a large saving response to the 1981 tax changes is hardly a prudent basis for sound public policy.

Further, despite the rhetoric that accompanied it, the 1981 tax bill contained few specifically targeted saving incentives. The reduction from 70 percent to 50 percent in the maximum marginal rate applicable to "unearned" income applies to relatively few taxpayers. The new IRA and expanded Keogh account provisions are potentially more important, but for many individuals they will only affect inframarginal rather than marginal saving flows (and hence, by the reasoning outlined above, will unambiguously *reduce* those individuals' total saving). By far the greatest part of the personal tax reduction enacted in 1981 and implemented during 1981–83 consisted of general across-the-board rate cuts, which are unlikely to have much impact on saving behavior even on the most optimistic rendering of the available evidence.

Finally, because the projected government deficits for the balance of the 1980s are so large, even a *doubling* of the personal saving rate, from the 1951–1980 average of 4.8 percent, to 9.6 percent, would merely finance the likely deficit, without leaving anything in addition to increase the economy's net capital formation. Only an extraordinary increase in personal saving, to magnitudes approximating that of several countries abroad, would meet both needs. Although the reasons why the U.S. personal saving rate is so low in comparison to that of some other countries remain imperfectly understood, they almost certainly involve some fairly fundamental aspects of societal arrangements rather than just marginal tax rates. Indeed, the simple cross-country correlation of growth rates and marginal tax rates is more often positive than negative.

Easy answer 2 is that a combination of large profits during the economic expansion and the liberalized depreciation allowances legislated in 1981 will provide business with a sufficient internal cash flow to boost the historically small corporate saving rate so as to enable business to undertake increased net capital formation without having to rely on personal saving raised in the capital markets.

The problem here is that there is no one-for-one correspondence between increased business cash flow and increased net investment. At the most immediate level, the typical corporation is likely to pay out some fraction of its increased after-tax cash flow, either by raising dividends or by buying shares in other corporations in the course of mergers and acquisitions. There has already been some evidence of such activity, although corporate dividend pay-outs usually respond to changes in cash flows only with a lag.

Further, even that part of the increased cash flow which corporations retain still does not bear any direct correspondence to overall business investment. As Table 11−3 shows, net private saving in the United States is quite stable as a share of GNP. This relative constancy suggests that, on balance, respective fluctuations in corporate saving and in personal saving about offset one another, as shareholders adjust their own direct saving for the saving that corporations do in their behalf. More sophisticated statistical analysis confirms this result, typically indicating a personal saving response great enough to offset more than one-half of whatever variations occur in corporate saving. The basic point is that what constrains net private investment (apart from government and foreign deficits) is net private saving, and rearranging the composition of net private saving is not the same as raising its total.

A closely related line of reasoning is that, because price inflation raises the effective tax rate on business investment, the recent slowing of U.S. inflation will reinforce the effect of liberalized depreciation allowances and hence boost business cash flows (and after-tax rates of return) still further. In fact, however, recent studies have shown that the overall effective marginal tax rate on income generated in the U.S. corporate sector is relatively insensitive to inflation. Inflation clearly distorts the *allocation* of capital by widening the dispersion among the different marginal tax rates applicable to different forms of corporate sector investment. At the average inflation rate prevailing in the United States during 1970−1979, for example, the range of marginal tax rates on income generated in the U.S. corporate sector extended from −105 percent (that is, a 105 percent subsidy) on machinery financed by sale of debt to pension funds and used in the commercial sector, to +111 percent (that is, more than full confiscation of proceeds) on buildings financed by sale of equity to households and used in commerce or industry. Nevertheless, inflation does not much affect the *overall* tax rate in a way that would plausibly reduce the *total amount* of capital formation.

Easy answer 3 is that the United States can finance both its government deficit and its net capital formation with foreign capital inflows, as investors abroad increasingly see both high U.S. interest rates and the stable American political and economic environment as strong attractions for their saving.

Further increases in foreign capital inflows (corresponding to a negative net foreign investment position in Table 11–3) may well occur. Moreover, such inflows do constitute a direct addition to the internally generated net saving available to finance both government deficits and private domestic investment. Why, then, are still further increases in foreign capital inflows not the most straightforward way to deal with the problem?

Especially in the context of concerns about the U.S. economy's international competitiveness, foreign capital inflows score points only by giving away the game. Foreigners hold an increasing amount of dollar assets (net of U.S. holdings of foreign currency assets) because Americans are buying more from abroad than foreigners are buying from the United States. The resulting imbalance of payments on current account inevitably makes foreigners net investors in dollar assets and, correspondingly, inevitably produces a negative U.S. net investment flow.

The direct counterpart to these capital inflows, therefore, is exactly the deterioration in U.S. net exports which has produced such widespread concern about American business competitiveness in recent years. The U.S. balance of merchandise trade has deteriorated from a seemingly invincible surplus before 1971, to a mixed pattern of relatively small surpluses and deficits during 1971–1976, to continual deficits of $25–35 billion per year during 1977–1982, to new record deficits of $69 billion in 1983 and $123 billion in 1984. This most recent decline in American competitiveness no doubt reflects in large part the real appreciation of the dollar's international exchange rate since 1980, which is itself a consequence of high real U.S. interest rates due to large budget deficits. Capital inflows in the range of 5 percent of GNP, as would be necessary to finance the entire federal deficit during the remainder of the 1980s, would imply merchandise trade deficits of $250–300 billion per annum throughout this period, probably an unacceptable solution from the perspective of those industries that would be principally affected.

Solving the government deficit problem with foreign capital inflows would therefore merely substitute a crowding out of the economy's foreign sector, via high real exchange rates, for the crowding out of the investment sector that would otherwise occur via high real interest rates. From the perspective of the economy's international competitiveness, that would hardly represent a satisfactory development.

In addition, if a policy relying on large foreign capital inflows is maintained for very long, it will sharply change still further aspects of the U.S. economy's international economic position. Net borrowing from foreigners is fundamentally different from borrowing from other Americans. Just two years of large net capital inflows have already entirely wiped out the United States' net international investment position accumulated gradually over nearly seven decades. Further inflows will increasingly make the United States a net debtor nation.

CHOICES FOR PUBLIC POLICY

In sum, under current tax and spending policies the United States will probably continue to enjoy an economic expansion in the short run, but will increasingly erode the foundation for its longer run prosperity: In the absence of more private capital formation than is consistent with today's enormous high employment federal deficit, gains in productivity—and hence ultimately in real standards of living—will be disappointing. Without significant correction in the dollar exchange rate, American business will not regain its international competitiveness. Massive reliance on foreign lending, driving the United States into net debtor status, will mortgage the nation's economic future.

What options are available, then, for preventing the U.S. government deficit in the 1980s from having the adverse effect on business capital formation, and the economy's productivity and competitiveness, that is likely under a continuation of the current fiscal policy? Any successful deficit reduction effort will no doubt require a willingness to compromise on other objectives also valued by important constituencies. In terms of the balance of saving and investment, what is required is, at the least, to reduce the deficit to 2 percent of GNP—in other words, below $100 billion per year. Deficit reduction of that magnitude would at least restore the availability of net private saving as it stood in the 1970s. Cutting the absorption of net private saving to its 1970s level would therefore require major changes, even in comparison with the Reagan administration's latest budget proposals, not to mention currently existing tax and spending legislation. The important question is *what* changes.

The standard trio of suggested ways to reduce the federal deficit in the medium-run future includes (1) cutting benefits under large entitlement programs, like Social Security or Medicare, (2) slowing the scheduled acceleration in defense spending, and (3) eliminating either the reduction in individual income tax rates which took effect at midyear 1983 or the indexation of the tax code that took effect at the beginning of 1985. The magnitude of the effort required to reduce the average deficit during the remainder of the 1980s to below $100 billion is such, however, that no one among these three steps would by itself be sufficient. Moreover, each of these three routes to deficit reduction carries potentially important negative side effects.

A policy designed to achieve meaningful deficit reduction while minimizing those side effects would therefore presumably combine at least two of these three elements, if not all three. By contrast, continuing to address the problem by relying only on the combination of economic growth and selected cuts in a variety of small spending programs would be a recipe for continuing large deficits, continuing high real interest rates and exchange rates, continuing weakness in business capital formation, and continuing failure in international economic competition.

12 THE GAP IN THE COST OF CAPITAL: CAUSES, EFFECTS, AND REMEDIES

George N. Hatsopoulos and Stephen H. Brooks *

In 1984 the U.S. merchandise trade deficit rose to a record $107 billion, or 3 percent of gross national product. This represents a major change from the surpluses recorded during most of the postwar period. Deterioration in the trade balance of manufactured goods was the major cause of this radical shift. Although much of the recent deficit increase can be attributed to the overvaluation of the dollar, longer term trends dating back to before the dollar's rise indicate a persistent loss of the competitive position of American manufacturers in world markets.

The decline in U.S. competitiveness is related to a broad spectrum of problems that have attracted attention in the growing literature on industrial competitiveness. These include a lack of focused support for research, slow commercialization of innovative technology, lack of cooperation between management and organized labor, fragmented

* The authors gratefully acknowledge the assistance of Dr. Elias P. Gyftopoulos, Ford Professor of Engineering, Massachusetts Institute of Technology, Dr. James M. Poterba, Assistant Professor of Economics, Massachusetts Institute of Technology, Dr. Lawrence H. Summers, Professor of Economics, Harvard University, and Mr. Thomas F. Widmer, Technology Advisor, Thermo Electron Corporation.

trade policy, slower growth in productivity of U.S. manufacturers, and too much emphasis on short-term results.

The common factor underlying most of these problems is the significant gap between the marginal cost of capital in the United States and in other industrialized nations, particularly Japan. In fact, over the last twenty-five years, American manufacturing firms have been paying three times as much for capital as have their Japanese counterparts. This has enabled Japanese manufacturers to invest two and a half times as much in tangible assets per worker as their U.S. competitors. Beyond this, they probably have invested a correspondingly greater amount than U.S. manufacturers in intangible assets[1] resulting from research and development activities, product improvement, and market development. Moreover, the much lower cost of funds in Japan makes the acceptable time horizon for investment decisions much longer than that for the United States. All these factors have contributed to the enormous difference in U.S. and Japanese rates of manufacturing productivity growth. Between 1973 and 1983 productivity grew by 6.7 percent per year in Japan compared to 1.8 percent in the United States.

The purpose of this chapter is to analyze the difference in the cost of capital for the manufacturing sector in the United States and Japan over the last twenty-five years, and to examine its causes, effects, and possible remedies. We have singled out Japan because over the past few years the impact of Japanese competition on U.S. manufacturers has been much greater than that of other foreign manufacturers. This is true both for the domestic U.S. market and for U.S. exports to world markets.

Although the cost-of-capital gap between the United States and Japan has persisted for over twenty-five years, its effects have only recently been recognized. At the end of the war, the United States started with a huge advantage over Japan, and all other nations as well. This initial advantage was due to our much stronger position in capital, technology, and know-how—an edge which has taken the Japanese almost forty years to overcome. Some erosion was deliberately planned by the United States in order to achieve a more stable and equitable world economy. The deliberate overvaluation of the dollar that lasted until the late 1960s was the principal tool in the effort. Over this entire period, the insidious effects of our chronically large cost-of-capital gap were not recognized by U.S. policymakers. These effects have taken decades to manifest themselves and are likely to require at least as long to undo.

The problem was exacerbated by two major surprises. The first was the energy crisis of the mid-1970s, which rendered obsolete a large part of fixed industrial capital. Large capital investments were required to improve energy productivity, but the high rates of inflation that followed the sharp rise in energy prices caused the cost of capital to rise at the most inappropriate time. The second factor was the decision to attack the high rate of inflation of the 1970s by means of a tight monetary policy designed to depress the economy, while simultaneously pursuing a very loose fiscal policy designed to stimulate the economy. The result was a deep recession, followed by a robust recovery and an overvaluation of the dollar.

Most of these gyrations can be corrected quickly if there is a will to do so. What will take much longer, even with an ideal set of corrective actions, is the restoration of the competitive position of our manufacturing industry which is vital to preserving and improving the standard of living of the American people.[2]

THE GAP

Defining the Cost of Capital

There is no broadly accepted definition for the term "cost of capital." It means different things to different people. In this chapter *cost of capital* is defined as the minimum, pre-tax, inflation-adjusted gross rate of return required to justify any incremental investment in an asset by a corporation. Such a return must cover depreciation of the asset, plus taxes to be paid on the return, less depreciation allowances and tax credits, plus required payments to investors and lenders supplying the funds used.

The cost of capital, evaluated at the margin, depends on the sources and uses of the incremental funds employed by corporations. We determine it in two steps. First, we calculate the marginal, after-tax, cost of funds, the incremental payments which must be committed by a firm to obtain each additional dollar of new investment. Then we integrate the marginal cost of funds with the taxes associated with the new investment to establish the cost of capital for each application of funds such as equipment, plant, and inventory purchases.

The concepts and techniques used here draw upon the previous work of one of the authors (Hatsopoulos 1983, 1984). This study differs from the previous work in that it focuses on large manufacturing corporations, that part of the corporate sector most directly involved in international competition. Calculations in the earlier studies were for the entire nonfinancial corporate sector. In addition, the current work extends the analysis of debt to include those funds, such as accounts payable, which are debts but which are interest free to the corporation using them. Finally, the current work relaxes the assumption used in one of the earlier studies that the marginal source of funds to rapidly growing Japanese corporations was entirely debt.

Sources of Funds

It is traditional to think of two sources of funds for corporations: equity and debt. In fact there are two distinct types of debt: interest-bearing debt, such as bank loans and corporate bonds; and interest-free debt such as payables, deferred taxes, and other deferred liabilities.[3] Each of the three categories (equity, interest-bearing debt, and interest-free debt) differs from the other two in exposure to risk, in taxation, and in cost.

Use of equity exposes a corporation to the least risk because it involves no fixed obligation to provide either returns or repayments. For the same reason, the supplier of equity funds is exposed to the biggest risk. Use of interest-bearing debt exposes a corporation to the biggest risk, and the supplier of funds to the least, because it involves a fixed obligation to provide returns and to repay the funds. Finally, the use of interest-free debt such as trade payables involves risks that range between the two extremes: fixed obligations to repay the funds but no obligations to pay returns.

The two sources of funds that require payment of returns (equity and interest-bearing debt) differ in the amount of taxation on these returns at the corporate level. The taxation usually corresponds to the risk classification of the payments—payments to equity holders are taxed, whereas payments to debt holders are not. This correspondence is strictly adhered to in the United States but not in Japan. For example, a U.S. financial instrument can be classified as debt for tax purposes only if the holders of the instrument are sufficiently insulated from the future performance of the company as to require, in substance as well as in form, a fixed commitment on the part of the company. Thus, it is very rare in the United States for the largest stockholder of a corporation also to be the largest debt holder, because such a situation would imply that the debt imposes no more risk to the corporation and no less risk to the provider than did the equity. In such a situation the U.S. Internal Revenue Service would disallow interest deductions. In Japan, the situation is quite different due both to the financial structure of their industry (see below) and their prevailing laws. As a result, the largest stockholders of a Japanese company are often the largest debt holders as well. (See Chase Financial Policy 1980: 7.6, 8.9.[4])

From the point of view of a corporation, the cost of interest-bearing debt is less than half that of equity funds for two reasons. First, because of differences in risk, debt holders require lower returns than do equity holders. Second, interest payments are tax deductible to the corporation, whereas the returns to equity holders are taxed. Interest-free debt, such as payables, constitutes the cheapest source of funds.

Financial Structure of Industry

For the most part, U.S. corporations operate independently of one another. Substantial ownership of the equity of a publicly traded corporation by another is rare. Moreover, equity holders are segregated from debt holders not only by virtue of tax rules but also because the rules of the Glass–Steagall Act prohibit banks from owning corporate equities.

In Japan, things are quite different. A large segment of industry consists of companies grouped around major banks. These groups are the modern version of the prewar *zaibatsus,* the large family-dominated industrial banking combines. They are commonly referred to as *keiretsus,* 'affiliations.' Moreover, the part of Japanese industry that competes most aggressively with the United States is targeted for direct or indirect government support.

The Japanese group structure provides major financial advantages to industry by making it possible for major equity holders also to be major lenders to corporations. Under these conditions, when a company needs external capital for expansion, both the company and the equity holder gain by having the needed funds supplied in the form of debt rather than equity because interest payments on debt are tax deductible. Thus, unlike the United States, growth companies in Japan are highly leveraged.

The banking system in Japan is financed principally by the enormous rate of personal savings, typically ranging from about 20 percent to 25 percent of disposable income (Blades and Sturm 1982) and by funds supplied by the Bank of Japan. Banks are instrumental in promoting personal savings. They conduct seminars and employ a large force of door-to-door salesmen to solicit the savings of individuals, especially employees of companies affiliated with the banks.

In the past decade, about 80 percent to 90 percent of all bank lending went to industry. Except for limited mortgage financing, consumer lending is practically nonexistent. This is partly due to the conditions imposed by the Bank of Japan on the use of funds it provides to commercial banks. In addition to funds provided by private banks, Japanese industry is also financed directly by government institutions, such as the Japan Development Bank, the Export-Import Bank of Japan, and the Small Business Finance Corporation. C.D. Elston (1981) estimates that direct government lending to industry has typically provided about 10 percent of total external funding.

Interest rates for Japanese industry have been quite stable, even during periods of high inflation such as occurred in the mid-1970s. In fact even when inflation rates were high, interest rates were kept low by government intervention. During those periods, consumer credit was virtually eliminated. In addition, interest rates and, as we shall see later, the cost of equity to industry have also been lower in Japan than in the United States. From this we infer that the risks inherent to industry are being partly assumed by the Japanese economy as a whole.

One of the vehicles used to distribute business risk to the economy as a whole is the government, operating through the Bank of Japan and the banking system. In addition, the Japanese government reduces business risks through the actions of the Ministry of International Trade and Industry. MITI frequently provides grants to industry which are repayable out of profits only if the project is successful. Elston estimates that as much as 30 percent of total corporate R&D expenditures are financed through MITI's Agency of Industrial Science and Technology.

We do not believe that government involvement in financing industry results in a more efficient spreading of risk compared with an efficiently running stock market. What the Japanese government does, however, is to spread business risk even among people who normally would be unwilling to accept it, such as the public at large. Whether or not such a system is fair and socially desirable, the fact remains that it enhances the international competitiveness of Japanese industry.

Sources of Funds at the Margin

The most expensive source of corporate funds is equity. Yet more than 50 percent of funds used by the great majority of U.S. corporations are equity funds. Japanese firms use proportionately much less equity, although the percentage has been increasing in the last few years. Table 12−1 shows a breakdown of the funds used by principal manufacturing corporations in the United States and Japan between 1977 and 1982.

The sources of interest-free credit to U.S. corporations are trade payables less trade receivables, taxes payable, and other deferred liabilities. In addition to those sources, Japanese companies avail themselves of special reserves for employee pensions and other contingent liabilities which can be deducted from taxable income. Obviously, these funds are limited by factors beyond the control of corporations.

Corporations do have much greater control over the amount of interest-bearing debt. Nevertheless, unlimited use of debt is not possible. Though it saves taxes, debt financing raises the probability of insolvency, thereby imposing a cost. An optimum ratio of equity to total liabilities—the equity ratio—would be one which just balances the marginal costs and benefits of additional debt. This optimum ratio minimizes the company's cost of funds and depends critically on the institutional and legal structure of the corporate environment.

The optimum equity ratio also depends on the particular nature of a business. For example, regulated monopolies such as utilities require less equity in proportion to their assets than do manufacturing companies. Moreover, the ratio depends also on the structure of ownership. For example, if both equity and debt are provided by one person or organization which assumes all business risks regardless of the equity ratio, then all new capital should be in the form of debt in order to minimize the company's taxes.

The actual equity ratio of any particular company varies significantly with time for several reasons. A company may be accumulating unusual amounts of equity in anticipation of making major investments or experiencing unfavorable economic times. Conversely, it may be accumulating unusual amounts of debt because of financial difficulties or in anticipation of issuing equity at more favorable terms in the future.

In the United States, the actual equity ratio of a whole business sector is relatively constant with time and constitutes a good measure of the optimum equity ratio, except when overall economic activity is unusually brisk or unusually depressed. In Japan the situation is quite different. Their financial system appears to dictate an optimum equity ratio which is two to three times smaller than in the United States. As

Table 12–1. Distribution of Funds Used by Principal Manufacturing Corporations, 1977–1982[a] (%).

Year	Net Interest-Free Debt		Net Interest-Bearing Debt		Equity	
	United States	Japan	United States	Japan	United States	Japan
1977	16.3	31.3	13.4	34.8	70.3	33.9
1978	16.6	31.7	13.9	32.9	69.6	35.4
1979	17.2	32.7	14.9	31.1	67.9	36.1
1980	17.3	33.7	15.9	29.4	66.8	36.9
1981	17.6	32.7	16.9	28.3	65.5	39.1
1982	19.0	31.3	16.5	28.5	64.5	40.2
1977–1979 Average	16.7	31.9	14.1	32.9	69.3	35.1

a. For details see Appendix 12A.

long as opportunities exist, a low equity ratio allows an industry to sustain extremely high rates of growth without equity financing. However, once a Japanese industry reaches the limits of its market opportunities and its growth slows, its investment needs decline and it has to find other uses for its profits. Increasing dividends is disadvantageous to the stockholders because the tax rate on dividends received is high. Unlike the United States, Japan prohibits the repurchasing of shares. For a corporation in this position, reducing debt is the only available option. This is exactly what has been happening during the last few years in Japan. Retained earnings are being used by many corporations to reduce debt. Thus, for these corporations, at least for the short term, the marginal source of funds is debt.

For our calculation of the cost of funds at the margin, we use the average balance-sheet ratios that prevailed between 1977 and 1979. We selected that period because there were no unusual economic strains in either country, and growth opportunities for most Japanese manufacturers had not yet saturated.

Cost of Debt

The nominal pre-tax cost of debt is determined by prevailing interest rates. In the United States, the AAA bond yield is the most appropriate for long-term trend studies because it is least influenced by short-term variations in the supply and demand for funds, and least subject to changes in classification practices.

In Japan, the interest rate is difficult to estimate because corporate borrowers are frequently required to hold compensating balances which result in an added cost. The measurement of the effective interest rate is further complicated by variations over time in the compensating balance requirements and rates. To avoid these problems, we take the yield on Nippon Telegraph and Telephone bonds to be the effective cost of debt. These interest-bearing bonds are essentially equivalent to AAA bonds in the United States. During much of the last decade, their return has been slightly above that of other industrial bonds, a finding supporting the view that compensating balance requirements change the effective interest rate. These assumptions consistently understate the cost of debt in both countries, but the error is relatively small. Table 12−2 shows the interest rates in the United States and Japan for years 1970 to 1984. Also shown are the inflation rates that prevailed in each country.

In order to calculate the cost of capital, we must express the cost of debt on a real after-tax basis. The calculations require knowledge of the statutory tax rates on corporate income and the expected rates of inflation. Tax rates are discussed in Appendix 12B and expected rates of inflation in Appendix 12C. Results for 1983 are shown in Table 12−3.

Cost of Equity

The cost of equity to corporations is the most esoteric concept used to calculate the cost of capital employed by businesses. The owner of a corporate share has a claim upon future cash flows in the form of dividends, capital distributions, and sale of the business. The underlying worth of a corporate share is, therefore, the discounted present value of such a stream of cash flows. If the stock market is efficient, stock prices will reflect the underlying value of the shares. Under this condition the present value of an expected cash stream will also equal the present value of an expected cash stream to an investor who buys

Table 12–2. Interest and Inflation Rates (%).

	United States		Japan	
	Yield on AAA Bonds[a]	Actual Inflation Rate[b]	Yield on NT & T Bonds[c]	Actual Inflation Rate[d]
1970	8.0	4.7	9.2	7.6
1971	7.4	4.2	7.9	6.6
1972	7.2	3.7	6.6	5.4
1973	7.4	5.6	8.5	10.5
1974	8.6	10.1	11.7	21.7
1975	8.8	7.7	9.6	11.6
1976	8.4	5.2	9.0	8.9
1977	8.0	5.8	7.4	7.1
1978	8.7	7.0	6.8	4.6
1979	9.6	9.0	8.7	3.5
1980	11.9	10.2	8.5	6.8
1981	14.2	8.7	8.3	4.8
1982	13.8	5.9	8.3	2.8
1983	12.0	3.7	8.0	1.9
1984	12.7	3.2	7.9	2.2

Sources:

a. Board of Governors of the Federal Reserve System
b. NIPA (1981, 1982, and 1984)
c. Bank of Japan
d. OECD (1984)

shares today, receives dividends for a time, and then sells the shares.

Each individual stockholder has his own perception of future cash flows, and his own discount rate for a given corporation. The combined effects of all such perceptions and rates result in the market value of the stock.

For the purposes of this study, we define the *cost of equity* as the rate at which expected future dividends, as perceived by a firm's managers, must be discounted in order to arrive at a present value equal to the market value of the firm's stock.

This definition recognizes explicitly the alternatives available to management when considering any potential investment opportunity.

Table 12–3. Corporate Cost of Debt in 1983 (%).

	United States	Japan
Nominal pre-tax cost at annual rate	12.0	8.0
Nominal pre-tax cost at instantaneous rate[a]	11.4	7.7
Statutory rate on corporate income[b]	50.3	54.7
Nominal cost after taxes	5.7	3.5
Projected rate of inflation	7.1	4.0
Real cost after taxes	−1.5	−0.5

a. An instantaneous rate compounded continuously for one year yields the annual rate. Unless othewise specified all rates shown in this chapter are instantaneous and in percent per year.

b. Includes state and local taxes.

The task of management is to maximize the value of corporate shares. In issuing additional shares to finance a new investment, management must weigh the dilution in value of existing shares against any expected increase in value resulting from the proposed investment. In other words, management should compare the incremental cost of new equity with the incremental return on such equity.

The perceptions of future corporate cash flows by investors, combined with their required return on equity, determine stock prices. Stock prices combined with the perceptions of management, which may differ from those of investors, determine the cost of equity for the company. In finance theory, it is customary to equate the rate of return required by equity investors to the cost of equity. This is so if the perceptions of future corporate cash flows by investors are the same as those of management. Perceptions of management, however, may differ substantially from those of investors because the two groups frequently possess different information. Since investment decisions are made by management responsible to maximize stockholders' wealth, the cost of equity that governs these decisions may differ from the required return on equity. In the United States, such a divergence probably occurred on a large scale in the mid-1970s. Modigliani and Cohn (1979) have suggested that in that period investors may have committed major errors in evaluating common stocks in the presence of high rates of inflation.

Equity is also added to a firm through retained earnings. For U.S. firms, retained earnings cost the same as new equity[5] because management has the option of repurchasing the firm's shares at current stock prices in lieu of investing in the company's business operations. Stock repurchases reduce the company's equity by an amount equal to the market value of the repurchased shares and thus return to the company an after-tax cash flow equal to the corresponding stream of expected cash flows. Forgoing the repurchase in favor of any other investment will, therefore, cost the company a return exactly equal to the discount rate of future cash flows—that is, the cost of equity.

Since Japanese firms are not allowed to repurchase their shares, they can only use earnings to pay dividends or repay debt.[6] This restriction, coupled with the high rate of taxation on dividends received by stockholders, reduces the cost of retained earnings below the cost of new equity. Because analysis of this effect is very complex, we have disregarded it. As a result, we overstate the cost of equity from retained earnings in Japan.

From a standpoint external to a corporation, the corporate after-tax cost of equity and the corporate pre-tax cost of debt are conceptually comparable, because both represent the pre-tax return to investors. On the other hand, from a standpoint internal to a corporation, the corporate after-tax cost of equity and the corporate after-tax cost of debt are conceptually comparable because both represent the after-tax cost to the corporation.

We derive the cost of equity by means of the procedure described in Appendix 12D. Briefly, aggregate dividends paid by corporations in any one year are cyclically adjusted to reflect a sustainable level of dividend payments. The projected growth rate of dividends is assumed to be the annual growth rate of dividends that can be sustained by the corporate sector in a steady state by means of retained earnings and a proportional amount of debt.

The cost of equity is determined as the discount rate which yields a present value of projected dividends equal to the corresponding aggregate market value of corporate equities. The data we use are for the whole corporate sector of either the United States or Japan. We assume that stock markets in each country are efficient. It follows that the calculated cost of equity applies to the manufacturing sector as well. As shown in Appendix 12D, the cost of equity of the corporate sector equals the sum of the average dividend yield adjusted for cyclical effects and dividend deductibility, plus the projected growth of the sector adjusted for new equity issues. For 1983, the results are shown in Table 12–4. For that year, as well as for all of the previous years studied, the cost of equity in Japan was much lower than in the United States. One manifestation of that fact is the much higher price-to-earnings ratios enjoyed by Japanese companies.

In the United States, dividends paid by a corporation are not deductible from the corporation's taxable income. Therefore, the corporation's cost of equity after taxes and the return to equity holders before taxes are the same. In Japan, however, 20 percent of the dividends are deductible. As a result, the cost of equity after taxes is less than the pre-tax return to equity holders.

The real pre-tax return to equity holders and the real pre-tax return to debt holders are listed in Table 12–5. In both the United States and Japan, the return to equity holders is greater than the return to debt holders, reflecting the risk premium that is inherent to equity. Both these rates of return, however, are significantly lower in Japan than in the United States. A large part of the difference is due, at least historically, to the combination of high savings rates and restrictions on capital flows that prevailed in Japan.

The last row of Table 12–5 shows the equity risk premium in 1983—the difference between the return to equity and debt holders. The equity premium is lower in Japan than in the United States. This result is surprising, especially in view of the fact that Japanese companies are much more leveraged than their American counterparts. Large leverage raises the risk of insolvency and reduces the equity basis across which this risk must be spread. In Japan, however, the risk of insolvency of major corporations is much lower than in the United States because of government involvement. Moreover, because of the interrelation of debt and equity in Japan, risk is spread more equally among both equity and debt holders. Finally, returns to equity holders are subject to lower rates of taxation than returns to debt holders because capital gains on equities, a major part of equity returns, are not taxed in Japan, and dividends received by individuals are taxed less than ordinary income (Shoven and Tachibanaki 1985).

Cost of Funds

We calculate the nominal after-tax cost of funds as the average of the nominal after-tax cost of equity and interest-bearing debt, weighted in the proportions shown in Table 12–1. The nominal cost of interest-

Table 12–4. Corporate Cost of Equity and Its Components, 1983 (%).

	United States	Japan
(1) Actual dividend yield	4.4	1.3
(2) Cyclically adjusted yield	5.5	1.4
(3) Tax-deductible fraction of dividend	0	19.1
(4) Yield, adjusted for dividend deductibility	5.5	1.2
(5) Projected real growth per share	3.1	4.0
(6) Real cost of equity after taxes, (4) + (5)	8.6	5.2
(7) Projected rate of inflation	7.1	4.0
(8) Nominal cost of equity, (6) + (7)	15.7	9.2

Table 12–5. Pre-tax Returns to Equity and Debt Holders, 1983 (%).

	United States	Japan
Real return to equity holders	8.6	5.4
Real return to debt holders	4.3	3.7
Equity risk premium	4.3	1.7

free debt is assumed to be zero. From 1970 to 1984, results of the calculations are shown in Table 12–6. It is seen that throughout the fifteen-year period, Japanese manufacturers have enjoyed a clear advantage.

Table 12–6. Nominal Cost of Funds in Manufacturing, 1970–1984 (%).

Year	United States	Japan
1970	7.40	6.04
1971	7.37	5.66
1972	7.14	5.20
1973	7.83	5.98
1974	10.40	7.78
1975	10.96	7.04
1976	10.16	6.60
1977	10.30	5.86
1978	11.00	5.30
1979	11.90	5.36
1980	12.24	5.21
1981	13.12	4.95
1982	13.05	4.71
1983	11.67	4.39
1984	9.06	4.23

Real Cost of Capital

Corporate funds are invested in four major categories of tangible assets: equipment, structures, land, and inventories. All these categories are accounted for in the corporate balance sheets.

In addition to investments in tangible assets, however, corporations make many other investments in intangible assets that are important to their future. These include investments in technology, product design and marketing, and service capabilities. Although many of these investments have lasting value, none is capitalized on an accounting basis and, therefore, none appears on corporate balance sheets. The sale of a company at a multiple of its book value, a common occurrence, frequently reflects the existence of these "intangible" assets.

The use of an asset in a company's operation involves a cost that is distributed over time. If this cost is distributed in proportion to the current replacement value of the asset, it is called the *service cost* of the asset. The present value of the service cost incurred over the life of an asset, discounted by the cost of funds, must equal the purchase price of the asset.

The *cost-of-capital services* or, simply, the *cost of capital* is defined as the ratio of the service cost incurred over a year divided by the average replacement value of the asset in that year. Cost of capital can be regarded as the rental rate a corporation must pay to cover the cost of funds used, the depreciation of the asset, and the taxes incurred. To evaluate the cost of capital, we use an extension of the approach developed by Hall and Jorgenson (1967). We account for inflation and the timing of tax payments and tax credits, and include intangible assets such as the technology resulting from an investment in R&D. Details about the cost of intangible capital are discussed further in Appendix 12F. For equipment and structures, the tax data and decay rates used appear in Appendixes 12B and 12G, respectively. For land we assume no depreciation. For inventories we assume no obsolescence[7] and no taxes on appreciation.[8] The results for 1983 are shown in Table 12–7. The fifth row shows the overall cost of tangible capital, which is the average of the costs of equipment, structures, land, and inventories, weighted by their relative use in the United States.

In addition to the cost of capital for tangible assets, Table 12–7 includes the real cost of a hypothetical investment in a venture that would take ten years to develop. A ten-year lag between investment and exploitation is not uncommon for major innovations.

The overall real costs of tangible capital to U.S. and Japanese manufacturing industries for the years 1961 to 1984 are shown in Figure 12–1. It can be seen from these graphs that the Japanese have enjoyed an enormous capital cost advantage over the entire period under consideration. This advantage became the largest during the mid-1970s. More recently, Japanese costs have increased and U.S. costs have tended to decline. Nevertheless, in 1984 the overall U.S. cost of capital including depreciation—the gross cost—was still more than 50 percent greater than in Japan.

The costs shown in Figure 12–1 and Table 12–7 include all of the costs associated with the use of an asset: the real cost of debt and equity, costs associated with taxes and tax credits, and the cost associated with the economic depreciation of the asset. With the exception of depreciation, all of these costs depend on the economic and financial environment in which a business operates. Depreciation, on the other hand, depends on the physical nature of the asset. To focus attention specifically on the effects of the economic environment, it is useful to exclude depreciation from the cost-of-capital calculations. We call this result the *net cost of capital*—the cost of capital excluding depreciation.

Table 12–7. Cost of Capital (Including Depreciation) in 1983 (%).

	United States	Japan	Shares of U.S. Corporate Tangible Capital
Equipment	17.3	14.4	32.6
Structures	16.4	9.3	30.1
Land	9.0	0.9	15.1
Inventories	9.0	0.9	22.2
Overall	14.0	8.0	100.0
Ten-Year Project	12.2	5.6	—

Figure 12–2 shows the real net costs of tangible capital in the United States and Japan over the period 1961 to 1984. The basic patterns of the net costs are similar to those in Figure 12–1. Of course, the striking divergence between the net costs in the United States and Japan during the mid-1970s is the same as that between gross capital costs. This mid-1970s divergence reflects the very different reactions in the United States and Japan to the sharp increase in energy costs after the first oil shock. In Japan, where capital markets are tightly controlled, the sharp increase in domestic inflation rates was not allowed to be translated into an increase in nominal corporate borrowing costs. As a consequence, the real cost of debt fell sharply during this period. In addition, the real cost of equity was virtually unchanged. On balance, the cost of funds fell sharply just at a time when the demand for new energy-enhancing capital investments rose sharply. Japanese policymakers clearly chose to keep capital costs low in order to spur these needed investments.

The contrast with the United States during this same period is profound. While the increase in inflation rates was not fully reflected in nominal corporate borrowing rates, it was to a much greater degree than in Japan. Thus, while the real cost of debt did decline somewhat in the United States, it was nothing like the decline experienced by Japan. In addition, the extraordinarily poor showing by the U.S. stock market drove up the yields on equity and, therefore, the cost of equity. Since equity is much more heavily used in the United States than in Japan, the effect of the increase in the real cost of equity overwhelmed the minor improvement seen in the real cost of debt. As a consequence, the real cost of capital rose just when it fell sharply in Japan. The

Figure 12–1. Real Cost of Tangible Capital Including Depreciation (%).

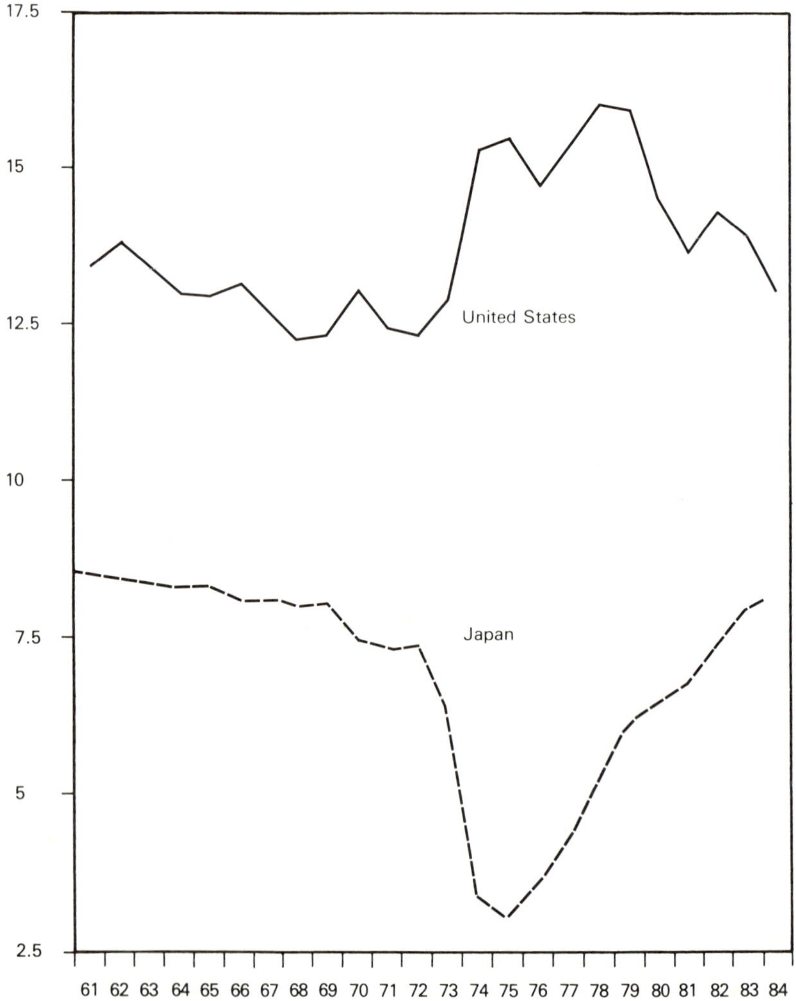

Figure 12–2. Real Net Cost of Tangible Capital Excluding Depreciation (%).

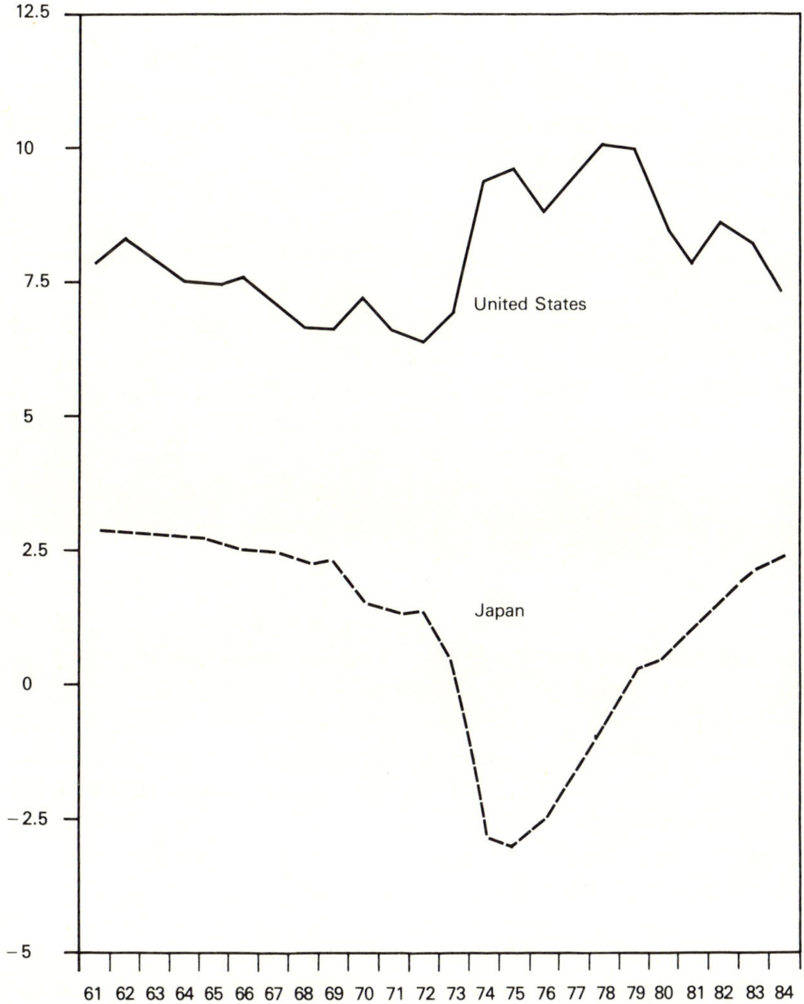

reduction in Japanese capital costs resulted in a strong spurt of investment during the second half of the 1970s that greatly increased the growth in capital per worker in Japanese manufacturing compared with the United States.

The recent increase in Japanese capital costs is consistent with the current economic situation in Japan. The rate of capital formation experienced during the late 1970s and early 1980s is no longer warranted. With huge export surpluses, the Japanese are forced to seek investment opportunities overseas. By raising the cost of capital, domestic industrial demand for funds has fallen, allowing this capital to flow overseas. The cautious opening of their capital markets is a further indicator of this desire. Investment opportunities within Japan are, at least temporarily, limited. However, we do not know whether this narrowing of capital costs between the United States and Japan will continue indefinitely.

Though smaller than its 1974 peak, the present cost-of-capital difference between the United States and Japan is still very large. Like differences in labor costs, differences in capital costs influence relative costs of goods traded on world markets. In addition, they result in higher investment rates within Japan. The larger and newer capital per worker ultimately translates into greater worker productivity, increased international competitiveness for manufacturers, and higher standards of living for all Japanese workers.

Labor-cost differences between the United States and Japan have steadily narrowed over the last twenty years. Between 1973 and 1983 real hourly compensation rates rose at an average annual rate of just 0.5 percent in the United States compared with 1.9 percent or nearly four times more in Japan. But with the high rates of investment, spurred by the very low capital costs, worker productivity in Japan has more than kept pace. In spite of the very high rates of wage gains, unit labor costs are declining faster in Japan than in the United States. Thus, real unit labor costs fell at an average rate of 4.5 percent per year in Japan compared with a 1.3 percent average annual rate of decline in the United States.

Figure 12—3 shows real net costs of capital—costs of capital excluding depreciation—for various types of assets in the United States and in Japan for 1983. The figure indicates that the present economic environment in the United States tends to favor investments in equipment and to discourage longer term ventures such as R&D projects. The opposite effect prevails in Japan. The principal factors producing this bias in the United States are the Investment Tax Credit for equipment and the

Figure 12–3. The Real Net Cost of Capital Excluding Depreciation, 1983 (%).

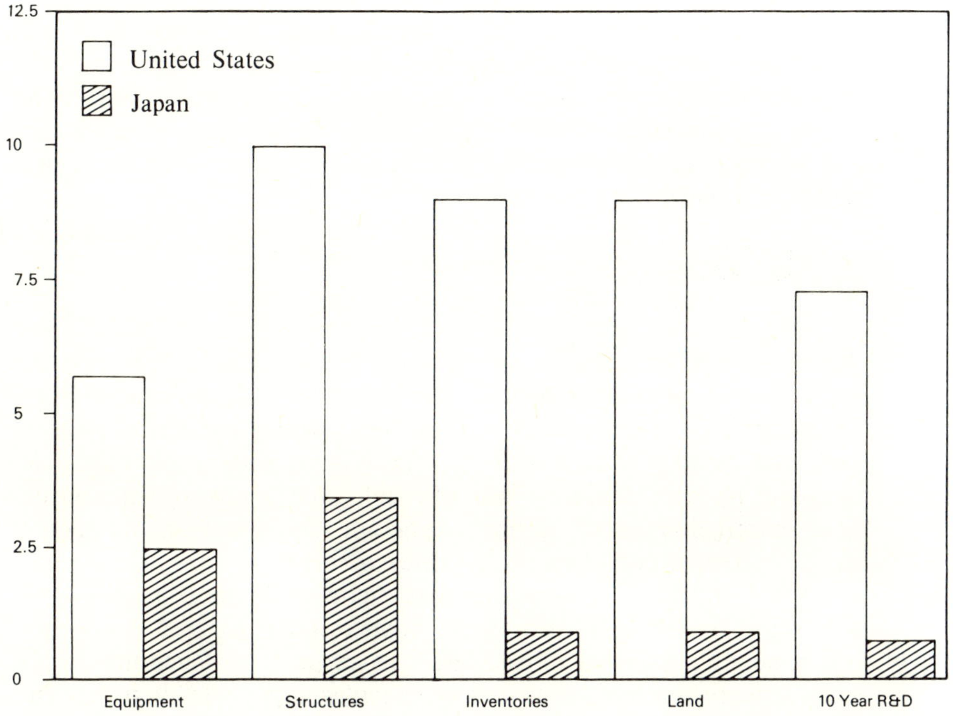

prevailing high real cost of funds. In Japan, the very low cost of funds makes ventures with long lead times much more attractive than in the United States. It is true that, in the United States, current R&D investments are expensed and that the R&D tax credit benefits investments in R&D. However, the contribution of these factors is small. The R&D tax credit applies to only a fraction of intangible expenses. Thus, this contribution does not offset the burdensome effects of the high cost of funds combined with long development lead times. The differences in net capital costs in the United States across alternative asset types have not always been as great as they are now. In fact, the economic environment prevailing in the early 1960s was almost neutral toward various types of assets and various segments of the manufacturing sector.

It is generally believed in America that free markets are more effective in creating wealth than any system involving central directions. Accordingly, it would be difficult to accept the proposition that a preferential tax treatment for certain assets could ever serve the national interest. On the other hand, let us not lose sight of the fact that biases in the costs of various capital services in the United States pale in the light of the differences between capital costs in the United States and Japan. For instance, in 1983, the net costs for equipment and for ten-year ventures in the United States were more than double and quadruple those in Japan, respectively.

America's Unique Advantage

The low cost of capital that is available to established Japanese manufacturers is not available to newly formed entrepreneurial firms. The latter have great difficulty in obtaining funds either as equity or debt. In contrast, many new American firms with a good idea are accorded, from time to time, virtually unlimited access to venture capital at extremely favorable terms. Although such booms in the venture capital market occur infrequently (they occurred for example in 1971 and 1983), relatively low-cost venture capital is far more readily available in the United States than in any other country.

It is nearly impossible to calculate an average cost of venture capital. Even though many new start-ups offer fantastic promise, very few survive. In mid-1983, many venture capital firms would calculate what they were willing to pay for a new start-up by discounting the projected future cash flow at extremely high after-tax rates of over 30 percent per year. If the probability of these firms succeeding were high, the cost of capital to the venture would be high, but if the probability of success proved to be low, the cost of capital to the venture would be extremely low.

Despite the difficulty in calculating a fair market value for the equity of new ventures, it is nevertheless generally accepted in the business community that venture capital investors are more willing to invest substantial amounts in the development of new ideas than are the managers of established corporations. Such an admission could be construed to mean that either private investors or business executives are irrational. It could mean, also, that there is a large fraction of the population throughout the world in general and in the United States in particular that is jackpot oriented. For such people, average expected rates of return are irrelevant; many people bet on lotteries that have well-documented negative expected rates of return. Therefore, even if one proved the rate of return of venture capital to be only half that for established companies, it is likely that plenty of willing investors would still be found. Nevertheless, we are convinced that the managements of publicly held companies tend to assign unreasonably high hurdle rates to new ventures, simply because the penalties for unexpectedly losing $1 million far outweigh the rewards of unexpectedly winning several million. The question still remains as to why venture capital is mostly confined to the United States. Perhaps the only answer to this is the uniqueness of the American culture.

THE CAUSES

The cost of capital is much lower in Japan than in the United States because the financial environment is more favorable to industry there than it is in America.

Some aspects of the Japanese tax code are more burdensome to business than in the United States. In particular, the tax rate on corporate income is higher in Japan. Moreover, there is no investment tax credit on equipment. Finally, depreciation allowances are somewhat lower than those currently provided in the United States. These tax disadvantages, however, are greatly outweighed by the advantages afforded by the financial system. This system provides industry with lower real interest rates, a lower real cost of equity, and access to a much greater proportion of investment funds with tax deductible returns than does the U.S. financial system at a comparable risk. Moreover, the higher corporate tax rate reduces the after-tax cost of debt. These factors result in a low marginal real cost of funds that overshadows the benefits of the lower U.S. corporate tax rates.

Underlying all of the factors contributing to the cost of funds differential between the two countries is the combination of the large saving rate intrinsic to the Japanese economy, and controls on the expatriation of funds. Over the past few years, the net saving rate in Japan has averaged 18 percent of gross domestic product as compared to only 5 percent of GDP in the United States. This enormous pool of savings and the historical acceptance of aggressive government intervention has provided the Japanese government with a wide latitude in controlling funds available for domestic investments. Moreover, it has been able to maintain both the cost of debt and the cost of equity at levels that best serve the objectives of the economy. In the United States, the Federal Reserve System through its controls of the domestic money supply can also indirectly affect interest rates and the cost of equity. But with net saving rates of only about 5 percent of GDP, this latitude is decidedly more limited.

Does the large consistent difference in the private savings rate for the two countries imply fundamental differences in national characteristics? We think not. Before the Second World War, U.S. and Japanese saving rates were generally similar. After the war, Japan deliberately attempted to boost the private rate of saving through government policies, laws, and private initiatives. Such national initiatives were just as deliberate and successful as, for instance, the U.S. effort to reduce traffic deaths to rates far lower than any nation in the world.

In contrast, from the mid-1930s to the late 1960s, America moved in the opposite direction from Japan. Consumption was promoted and investment deemphasized. Everyone participated in the process. The resulting consumption ethic that developed in America then was an appropriate response to excess production capacity. The situation, however, changed abruptly in the late 1960s when government consumption, stimulated by the combined effects of spending on the Johnson administration's Great Society programs and the Vietnam War outstripped the production capacity of the United States. Efforts to change the direction of the national priorities began in the early 1970s but, unlike Japan, the United States was unable to achieve quickly a lasting consensus with regard to economic goals and priorities. The slow progress that was achieved during the 1970s was overwhelmed by additional investment demands resulting from world energy shortages and the emergence of formidable foreign industrial competitors.

The problem of industrial competitiveness was aggrevated by asymmetries in trade practices and by the industrial targeting implemented by some of our trading partners. Although these factors are important, the analysis in this paper does not take them into account. The cost-of-capital gap discussed here does not reflect the many specific interferences of Japanese government in the private sector, such as the subsidies provided by the Ministry of International Trade and Industry. It does reflect, however, the more general policies in Japan that support savings and investment, and the general policies and subsidies in the United States that support consumption.

Tax policy in Japan appears to be well coordinated with broad economic objectives. Moreover, it is unaltered over long periods of time, a fact that allows industry to optimize its plans for investment. There are several provisions in the Japanese tax code that significantly contribute to the lower marginal cost of funds to corporations. First, capital gains on transactions involving securities, such as corporate shares, are not taxed, whereas capital gains on any other transaction are taxed at full rates. This provision encourages investments in the corporate sector and discourages investments in nonproductive assets. The high price-to-earnings ratios in Japan, currently averaging 26 to 1, are partly due to this provision.[9] Second, roughly 20 percent of dividends paid by corporations are tax deductible. This provision further lowers the cost of equity. Third, and most important, is the tax code definition of interest-bearing debt, funds whose returns can be deducted from taxable income. Because this definition is much more permissive than in the United States, Japanese firms have access to low-cost funds on a much greater scale than do their American counterparts.

The combined effect of these three provisions is an overall tax rate on marginal returns to all forms of funds that is significantly lower than that in the United States despite Japan's higher tax rates on corporate income and stockholder receipts. An intriguing aspect of the Japanese system is that the overall rate of taxation starts out very low for companies which embark upon major investment and which initially produce small returns. As soon as a company becomes successful and its profitability increases, the effective tax rate on investment returns increases. This occurs either through the taxation of individual stockholders, if the company distributes its earnings, or through the direct taxation of the corporation, if it retains the earnings and reduces debt.

Thus, in periods of vigorous expansion, when revenues from the taxation of ordinary individual income rise, the taxation of investment income declines. The opposite happens in periods of lower economic growth. Accordingly, the marginal cost of capital in Japan is much lower than its average cost. This effect is analogous to that produced in the United States by the R&D tax credit—it significantly lowers the marginal R&D costs but not the average R&D costs.

In contrast to Japan, the U.S. tax code is a conglomeration of laws and regulations, many of which are contradictory in purpose. It is a cumbersome and ineffective stimulus to economic growth. Moreover, because of its volatility, it drives industry toward a short-term opportunistic use of resources, rather than investments for purposes aimed at meeting the long-term needs of the country.

THE EFFECTS

The cost of capital relative to the cost of labor is one of the primary factors affecting investment behavior. In this chapter, no specific attempt has been made to integrate these capital- and labor-cost estimates into a model of investment behavior. It is clear from even a cursory inspection of the data, however, that capital-cost differences have had an enormous effect on relative rates of investment in the United States and Japan.

Historically, the primary Japanese advantage has been lower labor costs. This advantage, however, has been eroding rapidly. Japanese wage rates have been rising much more rapidly than those in the United States. Between 1973 and 1983, real compensation rates in manufacturing rose at an average annual rate of less than 0.5 percent per year in the United States compared with nearly 2 percent per year in Japan. Meanwhile, the capital-cost advantage has, until recently, remained very large. As a consequence, Japan enjoys capital costs that remain low relative to the price of labor. By contrast, the United States has faced a persistently high ratio of capital to labor costs.

The Japanese retain significant advantages with respect to both the absolute cost of capital and the price of capital relative to the price of labor. These two low costs have produced a logical and expected outcome: Japan is investing considerably more per worker than is the United States. The high Japanese investment rates have contributed to the very rapid gains in manufacturing workers' productivity. As a consequence, the large real wage gains seen in Japan have not reduced the rate of decline in real unit labor costs.

Investment Rates

Some summary figures capture the broad dimension of the story. In the manufacturing sector, where international competition is the strongest, the rate of investment in capital per worker in Japan has been between 2 and 2.5 times that in the United States.

The rates of gross fixed capital formation per manufacturing worker in the United States and Japan for the period 1970–1982 are shown in

Figure 12–4. They are derived from data provided by OECD (1984). In these and all subsequent comparisons between Japan and the United States, Japanese yen have been converted to U.S. dollars by means of OECD's purchasing-power-parity exchange rate. As can be seen, the Japanese investment per worker has remained consistently above the U.S. level. The only apparent narrowing in the investment-per-worker gap was over a brief period between 1973 and 1976. Since then the gap has widened significantly.

Capital-Labor Ratio

The obvious implication of the higher rates of manufacturing investment is that the stock of capital in Japan is growing more rapidly than in the United States. Starting from a very distinct capital-per-worker disadvantage in 1970, Japan has made tremendous progress. Between 1970 and 1981, the rate of growth in gross capital per worker was 7.6 percent per year, over twice the 3.5 percent annual rate of gain in the United States.

Unfortunately, these data on capital stocks from OECD are limited. The Japanese series contains data only through 1981, and refers to gross rather than net (after depreciation) capital stocks. The classes of investments made in the two countries are different and, therefore, different rates of depreciation would apply to the gross stocks of capital. In addition, the average age of the gross stock is considerably higher in the United States than in Japan. Both of these factors imply that the pattern of net stocks of capital (equipment and structures) per worker in the two countries would look different from that of gross stocks. Because much more of the U.S. gross capital stock was put in place before the Japanese stock, the net stock in the United States is a smaller fraction of the gross stock than it is in Japan.

Limited data on net capital stocks are available on a historical-cost or book basis from balance sheet information provided by Japanese manufacturing companies. In the United States, historical-cost data are available both from the Department of Commerce and from individual manufacturing company balance sheets. Table 12–8 shows net tangible assets per worker in the United States and Japan in 1974 and 1982. The Japanese data have been adjusted by OECD's purchasing-power-parity yen-dollar exchange rate.

Japanese balance sheets show net fixed capital after accounting for accelerated tax depreciation allowances. By contrast, U.S. historical cost

Figure 12–4. Manufacturing Fixed Investment per Year per Employee (Thousands of 1975 dollars).

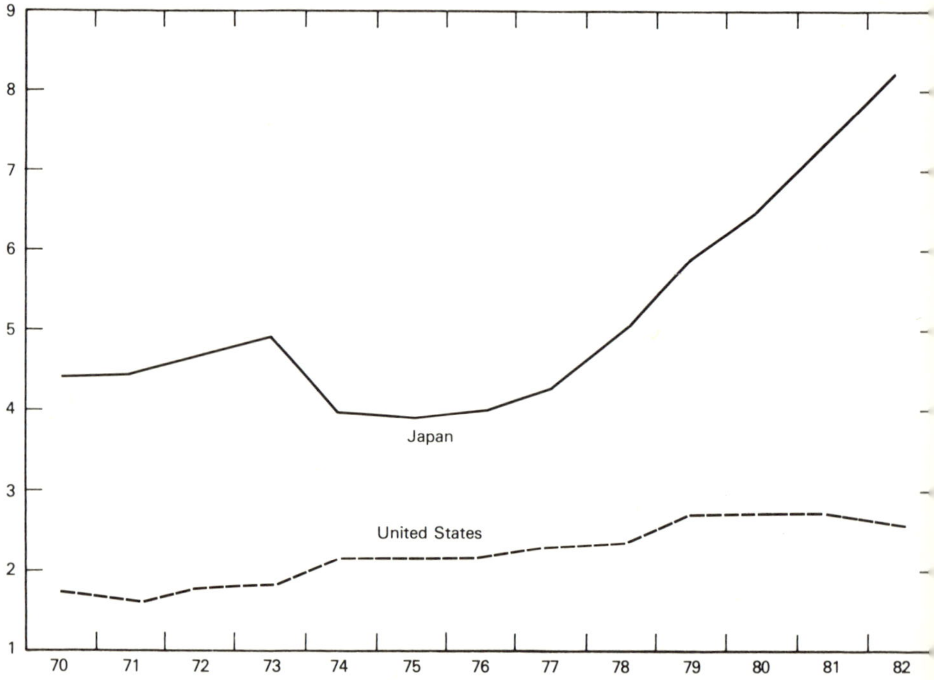

information on net fixed assets is based on straight-line depreciation. Because of this difference in accounting treatment, net fixed assets are underestimated in Japan more than in the United States. On the other hand, the data fail to account for increases in prices of fixed assets. This failure introduces another distortion. Here, however, the U.S. net fixed assets are underestimated more than the Japanese because inflation rates in the United States have been so much higher than in Japan.

We have not evaluated the relative effects of these two distortions. Nevertheless, it appears that net fixed assets per worker in Japan are considerably greater than in the United States. Thus, whether looking

Table 12–8. Net Tangible Capital per Worker, Principal Manufacturers (Historical Cost Concepts, Dollars per Worker).

	1974	1982
United States	12,778	32,143
Japan	20,423	48,099

Sources:

United States: *Value Line,* 400 Largest Manufacturers.

Japan: Principal Manufacturers from *Japan Statistical Yearbook,* adjusted by the OECD's purchasing-power-parity, yen-dollar exchange rate.

at gross or net capital stocks, the underlying story is clear: over the last thirty years the U.S. dominance in manufacturing capital-to-labor ratios, compared to the Japanese, has eroded significantly.

Productivity

One outcome of the higher rates of investment in Japan is that manufacturing worker productivity has been growing more rapidly than in the United States. Figure 12–5 shows the indices of output per worker hour. Between 1973 and 1983 Japanese manufacturing productivity rose at a rate of 6.7 percent per year compared with the U.S. rate of just 1.8 percent per year.

There is a solid body of analysis linking productivity with capital-labor ratios. On the basis of this link, it seems plausible to blame at least a portion of the differences in manufacturing productivity growth rates in Japan and the United States upon differences in capital formation in the two countries.

Capital formation alone cannot explain the entire difference. Most analyses of the U.S. productivity slowdown suggest that the drop in the rate of growth of capital per worker accounted for only a portion of the slowdown in U.S. productivity. By a similar reasoning, the differences in the rates of growth of capital per worker in the United States and Japan can explain only part of the differences in the growth rates of productivity.

Figure 12—5. Output per Hour in Manufacturing (1977 = 100).

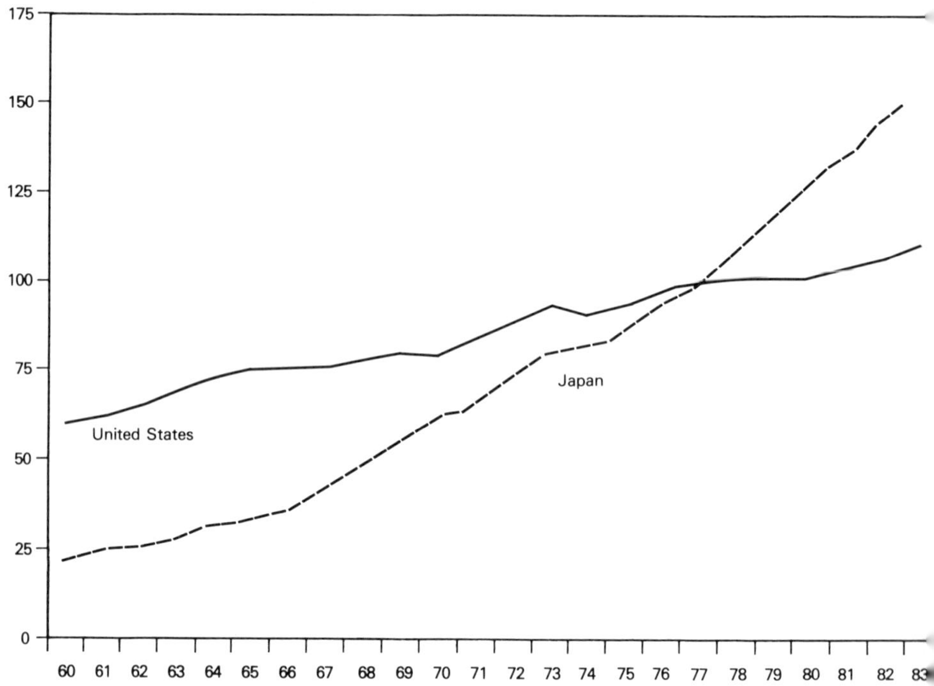

Source:
U.S. Department of Labor, Bureau of Labor Statistics

Intangible Investments

Intangible investments made directly by corporations in R&D, product design, and commercialization do affect productivity and go hand in hand with tangible investments. Yet they are typically ignored in traditional production-function analyses. Little is known about the relative rates of investment in intangibles in the United States and Japan, but there is reason to believe that the difference between the two rates may be very large.

Estimates of the capital-cost differential between the United States and Japan seem large enough to account for the observed differences in the rate of fixed investment in the two countries. With U.S. net capital costs for tangible assets running at roughly three times the level of Japanese capital costs, the rate of investment in Japan being twice that in the United States seems to fit.

Cost differences for intangible investments are even greater than for tangible or fixed capital in the United States and Japan. This is because intangible investments, such as in R&D, are often not expected to produce returns for many years. The higher the cost of funds (the rate at which returns from these investments should be discounted), the worse the cost of intangible investments. Because the biggest component of the capital cost difference between the United States and Japan is in the cost of funds it follows that intangible investments are much more costly in the United States than in Japan.

It may require five to ten years of R&D to introduce products based on new technologies. At the high-technology end of the spectrum, it is plain that the new very large-scale integration (VLSI) chips and products derived from biotechnology will have very long development times. Investments in R&D are undertaken only in an environment where costs (especially capital costs) are low enough to justify the wait. At the low-technology end of the spectrum, new products designed to maintain market share in maturing industries and investment in new marketing and distribution activities also require long gestation periods before any returns can be realistically expected.

We showed previously that the cost differentials for intangible assets are more extreme in the two countries than are the capital-cost differentials for tangible assets. It is this evidence that suggests that the rate of investment in intangibles is higher in Japan than in the United States.

It is difficult to estimate the relative size of intangible and tangible investments. Reasonable accounting detail is typically unavailable. Moreover, categorizing intangible expenses depends, to a certain extent, on the industry and company involved. A very rough estimate is that intangible investments run slightly more than tangible investments for a typical manufacturing firm. If Japanese companies invest roughly twice as much per worker as U.S. companies on tangible assets, then it is reasonable to assume that Japanese investments in intangible "knowledge" assets are at least twice as great as those of the United States.

Where should intangibles go in the production function? In *Accounting for United States Economic Growth 1929–1969,* Denison (1974) attributed most of the growth in productivity to "advances in knowledge." But these advances in knowledge, which are picked up as the residual in the typical production-function-based estimates of worker productivity, may not be so general and unspecified. The term *advances in knowledge* calls to mind the general education, skill, and experience of the population. In fact, the knowledge capital may be much more specific to a particular industry or company. Indeed, the residual in these production-function estimates is most surely capturing part of the increases in productivity due to intangible investments. Therefore, intangible investments must be contributing to increases in corporate profitability and productivity.

Detailed analyses of productivity trends are beyond the scope of this chapter. It seems clear, however, that the effect of intangible investments on manufacturing productivity are likely to be significant, and it is here that the Japanese have an even greater cost advantage than they do for tangible investments.

Long-term disparities in the rate of growth of productivity have lasting effects on both the international competitive position and the relative standards of living of the two countries. While much remains unexplained in the differences in productivity growth between the United States and Japan, the investment rate differences can easily be explained by lower costs of capital and a much larger pool of national savings.

THE REMEDIES

Some general points and principles on the design of an appropriate overall economic policy seem obvious. Today, the United States faces a potentially destabilizing mix of economic forces consisting of large federal deficits, low rates of personal saving, large trade deficits, a grossly overvalued dollar, and a very high cost of capital. This situation must be rectified. The essential requirements for long-term economic stability and growth are a rate of national savings consistent with the country's investment needs, a low rate of inflation, symmetric trade practices, and a cost-of-capital parity among trading partners.

Unfortunately, our current trade imbalance has diverted the policy debate away from the essential issues. The rapidly declining share held by U.S. industries in world markets (including the U.S. domestic market) has raised a major challenge to our historic commitment to free trade. It could result in harmful and counterproductive protectionist policies.

The tax system requires reform. The subsidies to consumption and penalties to saving must be removed to increase the national saving rate. Moreover, the federal budget must support rather than hinder this effort by moving quickly toward surplus. In addition, the current tax system is biased against some categories of capital income and in favor of others. These biases should be eliminated; the tax system should discriminate neither in favor of equipment nor against long-term ventures.

The most significant of the tax system's current capital-income biases is the double taxation of dividends. This unnecessarily boosts the cost of corporate funds and disproportionately raises the costs of long-term investments, both tangible and intangible. The double taxation of corporate investment can be prevented by making dividends paid by corporations tax deductible and by making capital gains from corporate equities nontaxable to investors. On the other side of the coin, present liquidation provisions, which result in the elimination of taxes on certain types of income and the capital gains exclusion for nonproductive investments should be curtailed.[10]

Although it is important to remove the pattern of biases and incentives in the current system, further efforts to reduce the cost of capital should be concentrated on lowering the cost of funds rather than lowering the costs of using specific kinds of investment goods. In contrast to targeted investment incentives, reductions in the cost of funds maintain neutrality across all investments, including intangible investments.

Sensible incremental changes in the U.S. tax system can help narrow the U.S.–Japanese differences, but it is unlikely that such remedies will completely eliminate the enormous Japanese advantage. At the heart of this advantage is the high Japanese saving rates. Until the United States is able to raise its saving rates, significant realignment of U.S. and Japanese costs of capital will probably be impossible. Clearly the first step is to bring the federal budget closer to balance. Beyond this, a more fundamental restructuring of the U.S. tax system may be required, to change the social and economic disincentives to save.

Although the problems facing our economy are staggering, there is a growing awareness of their nature and of the necessary corrections. Progress is being made in understanding the mechanisms and processes at work. We are confident that, as in the past, resources will soon be harnessed in a national effort to solve these problems.

APPENDIX 12A: CORPORATE BALANCE SHEETS

The balance sheets of principal U.S. and Japanese manufacturing firms were analyzed to determine the mix of net interest-free debt, net interest-bearing debt, and equity. The fraction of funds that are derived from each source is a critical factor in determining the cost of funds. Since the balance-sheet information is taken primarily from published sources from each country, and since each country has different accounting rules and procedures, certain adjustments were required to make the manufacturing balance-sheet information comparable. In comparing U.S. and Japanese balance sheets, we have tried to adopt U.S. accounting conventions. If Japanese balance sheets include items that would not ordinarily appear on U.S. balance sheets, then the Japanese data are altered to make them consistent with the United States. In addition to these changes designed to adjust for inter-country accounting differences, both sets of balance sheets had to be adjusted to make them consistent with relevant economic (as opposed to accounting) concepts. The purpose of the adjustments is to recast the "sources" or liability-and-net-worth side of corporate balance sheets so that the relative amount of funds from each of the three sources can be evaluated.

United States

Because we assume U.S. accounting conventions, the adjustment of U.S. balance sheets is somewhat less complicated than the adjustment of Japanese balance sheets. The data for the principal manufacturing companies in the United States were drawn from the 400 largest manufacturers (based on 1983 sales) in Value Line. Table 12A−1 shows the balance sheet information for 1979.

Net Interest-Free Debt. The interest-free debt used by a typical U.S. manufacturing company is the sum of net payables, and deferred taxes on revaluations.

Net Payables. Net payables are the sum of accounts payable, taxes payable, other current liabilities, deferred tax liability, and other long-term liabilities, plus other current liabilities, less the sum of accounts receivable, other current assets, and deferred assets (see Table 12A−2).

Table 12A–1. Financial Statement of Principal U.S. Manufacturers, 1979 ($ Millions).

Total assets	808,484	Total liabilities	431,628
Current assets	417,293	Current liabilities	239,268
Cash on hand and bank deposits	60,501	Accounts payable (inc. bills)	109,559
Accounts receivable (inc. bills)	154,541	Notes payable	30,723
Inventories	185,392	Taxes payable	30,488
Other current assets	16,859	Current portion of long-term	6,507
Long-term assets	386,337	Other current liabilities	61,991
Net plant	312,854	Long-term liabilities	161,529
Intangible	13,191	Minority-interest liability	8,819
Long-term investments	51,076	Long-term debt	127,245
Other long-term assets	9,216	Other long-term liabilities	25,465
Deferred assets	4,855	Deferred tax liability	30,833
		Net Worth	376,856

Source:
Value Line, 400 Largest Manufacturers.

Revaluations. The value of "net plant" is carried on U.S. balance sheets on an historical cost basis after tax depreciation. Because of inflation-induced increases in the replacement costs of these assets, their actual value is considerably greater than that shown on a corporation's books. The increase in value is, in effect, a source of funds that have been allocated to the "purchase" of fixed assets. If plant or machinery were sold, a portion of this increase in value would be taxed at the capital gains rate. The remainder would accrue to shareholders. Because the revaluations (or capital gains) on fixed assets are not taxed until they are realized, the corporation enjoys an interest-free loan from the government equal to their tax liability on the revaluations. This amount is added to net payables to calculate net interest-free debt.

Analysis of capital-stock information from the U.S Department of Commerce (1982a, b) revealed that for the manufacturing sector, the replacement value of net manufacturing equipment and structures was roughly 40 percent larger than the book value of these assets.[11] This 40 percent factor was used to calculate the revaluations of tangible assets. (Note that the implicit assumption is that the revaluation of land would have been roughly the same order of magnitude as that of

Table 12A–2. Net Payables, 1979 ($ Billions).

Payables	
Accounts payable	109.6
Taxes payable	30.5
Other current liabilities	62.0
Deferred tax liability	30.8
Other long-term liabilities	25.5
Less	
Accounts receivable	154.5
Other current assets	16.9
Deferred assets	4.9
Net payables	82.1

equipment and structures.) Thus total revaluations equaled $125.2 million (40 percent of net plant in 1979) of which $37.6 was added to net payables, and $87.6 to equity. The taxable portion of revaluations that were added to net payables was assumed to be taxed at 30 percent, the capital gains rate for corporations. Thus, the net sources of funds for principal U.S. manufacturing companies are calculated and shown in Table 12A–3.

Japan

Adjusting Japanese balance sheets is more complicated than adjusting U.S. balance sheets because of differences in accounting conventions. The balance sheet of the principal Japanese manufacturing companies for 1979 is shown in Table 12A–4. The data are drawn from the *Statistical Yearbook of 1982*, from the Statistics Bureau of the Office of the Prime Minister of Japan.

Net Interest-Free Credit. This category includes all of the sources of funds for which the corporation pays nothing. In Japan it is composed of three parts: net payables, deferred taxes on reserves, and deferred taxes on revaluations.

Table 12A−3. U.S. Principal Manufacturers: Sources of Funds, 1979.

	$ Billions	(%)
Net interest-free credit	119.7	17.2
Net payables	82.1	
Deferred taxes on revaluations	37.6	
Net interest-bearing debt	104.0	14.9
Debt	164.5	
Less cash and deposits	60.5	
Equity	473.4	67.9
Net worth (book)	377.0	
Minority interest	8.8	
Net revaluations	87.6	
Total	697.1	100

Net Payables. Net payables are equal to the sum of net accounts payable (including net bills payable), plus other current liabilities, less other current assets, less deferred assets. Derivation of the net payables total for 1979 is shown in Table 12A−5.

Reserves. In Japan, corporations are allowed to maintain tax-deductible reserves to a much larger extent than in the United States. These reserves are primarily of two types: long-term reserves and specific reserves. By far the greatest of these is the long-term reserves for deferred pension liabilities. In 1979, these reserves (listed in Table 12A−4 as "Other fixed liabilities") amounted to nearly 5 percent of all sources of funds. In the United States, corporations are not allowed to maintain as extensive reserve accounts for pension liabilities. In addition, "specific reserves," which are designed to cover specific contingencies are another source of funds that are not available to U.S. corporations to the same degree as Japanese corporations.

By not allowing U.S. corporations these deductions for reserve accounts, U.S. accounting and tax rules tacitly assume that they are not necessary. Therefore, to make the balance sheets of Japanese corporations consistent with those of U.S. corporations we considered both of these reserve categories to be economically unnecessary. The fact that reserves have already been deducted for tax purposes, when in fact they

Table 12A–4. Financial Statement of Principal Japanese Manufacturers, 1979 (10 Millions of Yen).

Total assets	7,308,232	Total liabilities and net worth	7,308,232
Current assets	4,511,851	Current liabilities	4,072,433
Cash on hand and bank deposits	741,629	Accounts payable (including bills)	1,523,412
Accounts receivable (including bills)	1,774,918	Short-term debt	1,323,963
Securities	387,050	Other current liabilities	1,225,059
Inventories	1,325,265	Fixed liabilities	1,733,453
Other current assets	282,989	Corporate debentures	306,159
Fixed assets	2,792,395	Long-term debt	1,087,278
Tangible	1,845,007	Other fixed liabilities	340,016
Intangible	16,442	Specific reserves	79,682
Investments and other assets	930,944	Net worth	1,422,667
Deferred assets	3,987		

Source:
Japan Statistical Yearbook 1982 (Statistics Bureau, Prime Minister's Office).

Table 12A–5. Net Payables, 1979 (10 Billions of Yen).

Payables	
Accounts payable	1,523
Other current liabilities	1,225
Less	
Accounts receivable	1,775
Other current assetts	283
Deferred assets	4
Net Payables	687

are unnecessary, means that the Japanese firms have been given an interest-free loan equal to the value of the reduction in taxes due to the deduction of reserves. That portion of reserves becomes a component of net interest-free debt. The remainder is included in equity. Net interest free credit is therefore increased by the product of the marginal

Table 12A−6. Total Reserves, 1979 (10 Billions of Yen)

Total	420
Other fixed liabilities	340
Specific reserves	80
Of which:	
Net interest-free credit	258
Equity	162

tax rate on manufacturing corporations of 0.6148 and the total amount of reserves. The remainder of reserves are added to net worth. Total reserves and their reallocation are shown in Table 12A−6.

Revaluations. The final category of interest-free credit is the deferred taxes on the revaluations in tangible assets. As noted in the discussion of U.S. balance sheets, if the firm were to be liquidated, shareholders would be taxed on the difference between replacement and book value. Only the after-tax portion of the revaluation should be counted as equity, the remainder amounts to an interest-free loan from the government.

Japanese balance sheets carry tangible assets at historical costs after tax depreciation (including accelerated depreciation). The actual value of these assets is considerably higher due to the increases in asset prices and due to the accelerated depreciation. The first step is to estimate the degree by which the actual replacement cost of the tangible assets exceeds the book value in 1979. Without any reliable aggregate information on Japanese net manufacturing capital stocks at replacement cost, this estimate is, at best, very tentative. In the United States the ratio of replacement to historical costs of manufacturing fixed capital is around 1.4. U.S. inflation rates for capital equipment consistently exceeded Japanese rates during the 1970s. Because of this, one would have expected that the Japanese ratio would be somewhat lower than 1.4. However, the prices of land in Japan have grown much more rapidly than in the United States (perhaps as much as twice as fast). This would imply a higher ratio depending on the overall importance of land on corporate balance sheets in the United States and Japan. The fact that tangible assets are net of tax depreciation suggests that the

Table 12A–7. Revaluations, 1979 (10 Billions of Yen).

Total	1,845
Interest-free credit portion	1,134
Equity portion	711

Table 12A–8. Japanese Principal Manufacturers: Sources of Funds, 1979.

	(10 Billions of Yen)	(%)
Net interest-free credit	2,079	32.7
Net payables	687	
Deferred taxes on reserves	258	
Deferred taxes on revaluations	1,134	
Net interest-bearing debt	1,975	31.1
Debt	2,411	
Corporate debentures	306	
Less cash	742	
Equity	2,296	36.1
Net worth (book)	1,423	
Net reserves	162	
Net revaluations	711	
Total sources of funds	6,350	100

replacement-book ratio should be considerably higher in Japan than in the United States since U.S. assets are net of straight-line depreciation. On balance we have assumed that the revaluation of tangible assets for the principal Japanese manufacturing companies was equal to 100 percent of their book value in 1979. These revaluations and their allocation to net interest-free debt and equity are shown in Table 12A–7.

Net Interest-Bearing Debt. Net interest-bearing debt is simply equal to the total amount of debt in the corporation (both short and long term), plus corporate debentures, less the amount of cash on hand and in banks.

Net Worth. Net worth equals the stated value of net worth from the balance sheets plus net reserves (net of the deferred taxes on reserves) plus net revaluations (net of the deferred taxes). Table 12A−8 shows the complete detail for Japan.

APPENDIX 12B: TAX DATA

United States

Tax parameters are listed in Table 12B−1. The average marginal state tax rate on corporate profits (column 2) is derived by averaging the statutory corporate tax rates in thirty-eight of the major industrialized states, weighting each tax rate by the state's value-added in manufacturing. The averaging is restricted to thirty-eight states because of data availability. The results are comparable to the measure based on the National Income and Product Accounts derived by dividing state and local corporate profits tax liabilities by before-tax profits in NIPA.[12] For example, in 1980 and 1981, our average is 8.0 percent versus the NIPA measure of 7.6 percent. Our series is preferred because it captures a marginal rate whereas the NIPA measure does not.

The statutory corporate income tax rate (column 3) is a combination of columns 1 and 2, and includes the deductibility of state taxes against federal taxes. This tax rate was used in evaluating the cost of funds and the cost of capital services.

Historically, effective corporate tax rates have been a lot lower than statutory tax rates. In 1979, for instance, the combined federal and state statutory corporate tax rate was 50.3 percent, whereas the effective rate (calculated as flow-of-funds tax accruals divided by before-tax book profits) was 36 percent. Some of this difference can be accounted for by a variety of tax credits. The largest of these, for domestic earnings, is the investment tax credit (ITC). In 1979 the total ITC claimed by all corporations was $14.6 billion. Even taking this into consideration, however, we find that statutory and effective corporate tax rates are still

Table 12B−1. U.S. Tax Code Parameters.

Year	(1)	(2)[a]	(3)[b]	(4)	(5)	(6)	(7)[c]	(8)[c]	(9)	(10)
1961	52.00	4.90	54.35	48.35	—	—	200	200	13	31
1962	52.00	5.00	54.40	48.40	7.00	7.00	200	200	12	30
1963	52.00	5.00	54.40	48.40	7.00	7.00	200	200	12	30
1964	50.00	5.00	52.50	46.50	7.00	—	200	200	12	30
1965	48.00	5.00	50.60	44.60	7.00	—	200	200	12	30
1966	48.00	5.10	50.65	44.65	—	—	200	150	12	30
1967	48.00	5.30	50.76	44.76	7.00	—	200	200	12	30
1968	48.00	5.60	50.91	44.91	7.00	—	200	200	12	30
1969	48.00	6.00	51.12	45.12	—	—	200	150	12	30
1970	48.00	6.30	51.28	45.28	—	—	200	150	12	30
1971	48.00	6.80	51.54	45.54	7.00	—	200	150	10	30
1972	48.00	7.20	51.74	45.74	7.00	—	200	150	10	30
1973	48.00	7.40	51.85	45.85	7.00	—	200	150	10	30
1974	48.00	7.50	51.90	45.90	7.00	—	200	150	10	30
1975	48.00	7.60	51.95	45.95	10.00	—	200	150	10	30
1976	48.00	7.70	52.00	46.00	10.00	—	200	150	10	30
1977	48.00	7.80	52.06	46.06	10.00	—	200	150	10	30
1978	48.00	7.90	52.11	46.11	10.00	—	200	150	10	30
1979	46.00	8.00	50.32	44.32	10.00	—	200	150	10	30
1980	46.00	8.00	50.32	44.32	10.00	—	200	150	10	30
1981	46.00	8.00	50.32	44.32	10.00	—	ACRS(e)	ACRS(s)	5	15
1982	46.00	8.00	50.32	44.32	10.00	5.00	ACRS(e)	ACRS(s)	5	15
1983	46.00	8.00	50.32	44.32	10.00	5.00	ACRS(e)	ACRS(s)	5	15
1984	46.00	8.00	50.32	44.32	10.00	5.00	ACRS(e)	ACRS(s)	5	15

Column 1: Statutory federal income tax rate (%)
Column 2: Estimated average statutory state income tax rate (%)
Column 3: Statutory corporate income tax rate at the margin (%)
Column 4: Effective corporate income tax rate at the margin (%)
Column 5: Total investment tax credit for equipment (%)
Column 6: Investment tax credit that reduces tax basis (%)
Column 7: Depreciation code for equipment
Column 8: Depreciation code for structures
Column 9: Average tax life for equipment (years)
Column 10: Average tax life for structures (years)

a. A state corporate income tax rate for the United States was computed by weighing the statutory marginal tax rates in thirty-eight states on the basis of each state's adjusted value-added by manufacture (as defined in the *Statistical Abstract of the United States*). The sample was restricted to thirty-eight states because data were unavailable for the remaining states. Data were available only for even-numbered years through 1978. All other years are estimated. The figures for odd-numbered years were obtained by interpolation. *Source: The Book of States* (Lexington, Ky.: The Council of State Governments).

b. Because state income taxes are a credit against federal taxes, the combined tax rate at the margin is

$$\text{Federal tax rate} + \text{state tax rate} \times (1 - \text{federal tax rate}).$$

c. 200 and 150 refer to double declining and 1.5 declining balance depreciation under Asset Depreciation Range. ACRS(e) and ACRS(s) stand for "Accelerated Cost Recovery System" applicable to equipment and structures, respectively.

very far apart. There are several reasons for this difference. First, several deductions are counted in the Internal Revenue Service definition of income, but not in the flow of funds. Loss carry forward and earnings of domestic international sales corporations (DISC) are two of the largest such deductions. In addition, dividends and capital gains are also treated differently in the flow of funds and by the IRS, resulting in further divergences between effective and statutory tax rates calculated from the NIPA data.

The effective tax rates shown in column 4 of Table 12B–1 include an adjustment to account for these differences between statutory and effective tax rates. Analysis revealed that, after taking account of the investment tax credit, effective rates were on average about six percentage points below statutory rates. Thus, for the purposes of projecting corporate cash flow for use in the sustainable growth calculations (Appendix 12E), statutory rates were reduced by six percentage points. In principle, the marginal dollar of corporate income is taxed at the peak statutory corporate rate. In practice, however, there are special deductions and credits that imply a much lower tax rate should be used in projecting cash flows. Had this not been taken into consideration, the estimates of projected growth would have been biased downward to levels that were much lower than actually experienced. As noted previously, however, the statutory tax rate, not the effective, was used in calculating the cost of capital services.

The effective tax rates do not include the federal surtax that was imposed from 1968 to 1971 because of the Vietnam War. We assumed that this surtax must have been perceived as a temporary abberation and, therefore, that it should not have affected estimates of long-term growth.

The reductions in tax basis for depreciation allowances for equipment are listed in column 6. The depreciation schedules that were assumed for equipment and structures are listed in columns 7 and 8, respectively.

The average service lives of equipment and structures in the nonresidential, nonfarm, private business sector are thirteen and thirty-one years, respectively (U.S. Department of Commerce 1982a, b).

Japan

Two aspects of the tax system are particularly relevant in any calculation of the cost of capital services. The first is the provision affecting interest deductibility and the second concerns investment credits and depreciation allowances. In Japan, interest payments may be deducted from taxable profits, just as in the United States. The situation is more complicated, however, because a corporation's marginal tax rate will vary depending upon its dividend-to-profits ratio. For instance, in 1983 the national tax on retained profits was 42 percent and on distributed profits 32 percent. Inhabitants' tax (τ_{INH}) is levied by localities at a rate (in 1983) of 17.3 percent of national taxes.[13] Finally, a deductible enterprise tax (τ_{ENT}) of 12.0 percent is levied on the firm's entire taxable profits.[14] Thus, the 1983 full marginal tax rate of the firm on retained earnings is

$$\tau_{RET} = A + B \times 0.42 = 0.547 \tag{12B.1}$$

where

$$A = \frac{\tau_{ENT}}{1 + \tau_{ENT}} = 0.107 \tag{12B.2}$$

and

$$B = (1 + \tau_{INH}) \times (1 - A) = 1.047 \ . \tag{12B.3}$$

The 1983 full marginal tax rate on dividends is

$$\tau_{DIV} = A + B \times 0.32 = 0.442 \ . \tag{12B.4}$$

This structure of tax rates is equivalent to one in which 19.1 percent of dividends paid are deductible for tax purposes.

If a fraction α of dividends is deductible, then

$$\tau_{DIV} = \tau_{RET}\,(1 - \alpha) \tag{12B.5}$$

and, therefore, in 1983

$$\alpha = 1 - \frac{\tau_{DIV}}{\tau_{RET}} = 0.191$$

A surtax on corporate income was instituted in 1984 for a period of three years. As in the case of the United States, this surtax should be

Table 12B–2. Japanese Corporate Tax Rates, (%).

Year	(1)	(2)	(3)	(4)	(5)	(6)	(7)
1961	12.0	13.9	26.0	37.0	48.3	37.2	23.1
1962	12.0	13.9	26.0	37.0	48.3	37.3	23.1
1963	12.0	13.9	26.0	37.0	48.3	37.3	23.1
1964	12.0	13.9	26.0	37.0	48.3	37.3	23.1
1965	12.0	13.9	26.0	37.0	48.3	37.3	23.1
1966	12.0	14.7	26.0	35.0	46.6	37.3	19.8
1967	12.0	14.7	26.0	35.0	46.6	37.3	19.8
1968	12.0	14.7	26.0	35.0	46.6	37.3	19.8
1969	12.0	14.7	26.0	35.0	46.6	37.3	19.8
1970	12.0	14.7	26.0	36.7	48.4	37.3	22.8
1971	12.0	14.7	26.0	36.7	48.4	37.3	22.8
1972	12.0	14.7	26.0	36.7	48.4	37.3	22.8
1973	12.0	14.7	26.0	36.7	48.4	37.3	22.8
1974	12.0	14.7	26.0	36.7	48.4	37.3	22.8
1975	12.0	17.3	30.0	40.0	52.6	42.1	19.9
1976	12.0	17.3	30.0	40.0	52.6	42.1	19.9
1977	12.0	17.3	30.0	40.0	52.6	42.1	19.9
1978	12.0	17.3	30.0	40.0	52.6	42.1	19.9
1979	12.0	17.3	30.0	40.0	52.6	42.1	19.9
1980	12.0	17.3	30.0	40.0	52.6	42.1	19.9
1981	12.0	17.3	32.0	42.0	54.7	44.2	19.1
1982	12.0	17.3	32.0	42.0	54.7	44.2	19.1
1983	12.0	17.3	32.0	42.0	54.7	44.2	19.1
1984	12.0	17.3	32.0	42.0	54.7	44.2	19.1

Column 1: Enterprise tax rate
Column 2: Inhabitants tax rate
Column 3: Basic tax rate on dividends
Column 4: Basic tax rate on retained earnings
Column 5: Full marginal tax rate on retained earnings
Column 6: Full marginal tax rate on dividends
Column 7: Implicit rate of dividend deductibility

perceived as a temporary abberation and, therefore, we have not included it in our calculations.

A list of corporate tax rates in Japan from 1960 to 1984 is given in Table 12B–2. Tax depreciation rules allow for assets costing less than 100,000 yen to be expensed; all other assets are eligible for depreciation according to either a straight-line or declining-balance method.

Table 12B—3. Average Tax Life and Depreciation Rates in Japan, (% per Year)

Asset Type	Average Tax Life (Years)	Straight Line Depreciation	Declining Balance Depreciation
Reinforced concrete building	65	1.6	3.5
Wood building	26	3.9	8.5
Factories	35	2.9	6.4
Machinery	11	9.0	18.9
Trucks	4	25.0	43.8

Most firms appear to choose declining-balance depreciation (Touche-Ross International 1980:95).

Useful lives for tax purposes and depreciation allowances for different asset types are listed in Table 12B—3.

There are few investment tax credits in Japan. Initial allowances have been provided in recent years for pollution control devices and for energy-saving technology. In addition, firms which can show that their assets will depreciate faster than the prescribed rates are able to obtain accelerated depreciation allowances. In our cost-of-capital calculations we have ignored these special incentives.

There is one further complication in calculating effective tax depreciation values. This concerns the depreciation allowed in the first year of the asset's life. If the asset is installed in the first half of the firm's accounting year, the firm may claim depreciation for the number of months for which it uses the asset. If the asset is installed at any time in the second half of the year, however, the firm may claim depreciation for one half year's asset use.

APPENDIX 12C: ESTIMATION OF EXPECTED INFLATION

One of the fundamental problems encountered in calculating the elements of the cost of capital, such as the real cost of funds and projected depreciation allowances, involves the estimate of the expected rate of inflation. Because there are widely differing views about future price changes among different individuals, an aggregate measure will, of necessity, seem inappropriate to some people.

One source of information about inflationary expectations is the bond market. If the nominal bond yield, i, is determined according to the principle that investors seek a certain real interest rate before taxes, i^*, in constant dollars, then the rate of expected inflation π^e would be[15]

$$\pi^e = i - i^* . \tag{12C.1}$$

But this equation is ambiguous (Feldstein and Summers 1978) because a high nominal interest rate may reflect either a high expected rate of inflation, or a high real interest rate, or a combination of these two factors.

There is some evidence that until the late 1970s, the real return on corporate bonds in the United States was relatively constant, at about 2.5 percent. Consequently, a fair approximation for expected inflation could be calculated by subtracting 2.5 percentage points from nominal interest rates. The result of this calculation is shown in column 2 of Table 12C−1. In the past few years, however, most indicators of inflationary expectations which came from outside the bond market would lead one to conclude that the real interest rate has in fact risen well beyond that value.

Column 1 of that table shows an expected inflation series based on the method of adaptive expectations which assumes that individuals learn, albeit slowly, by looking at past rates, and determine expected inflation in each year accordingly. We determined this series as a weighted average of the inflation during the year, n, and that of the preceding five years, the weight given by the relation

$$\textit{Weight in year } (n-k) = \frac{(0.7)^k}{\sum_{m=0}^{5} (0.7)^m}.$$

The inflation rate in any one year was based on the deflator for personal consumption expenditures (National Income and Product Accounts, 1981, 1982, and 1984).

Visual comparison of columns 1 and 2 shows smaller divergences in the 1960s and 1970s when monetary policy was more or less constant and larger divergences in the 1980s when monetary policy was much tighter.

Based on these considerations and other assessments, we concluded that the series shown in column 3 of Table 12C−1 calculated as the

Table 12C−1. Alternative Estimates of Inflation Expectations in the United States, (%).

Year	(1)	(2)	(3)	Year	(1)	(2)	(3)
1961	1.73	1.79	1.75	1973	4.54	4.71	4.60
1962	1.64	1.77	1.68	1974	6.31	5.75	6.13
1963	1.54	1.70	1.59	1975	6.75	5.99	6.50
1964	1.46	1.84	1.58	1976	6.27	5.63	6.06
1965	1.52	1.93	1.65	1977	6.13	5.25	5.84
1966	1.97	2.53	2.15	1978	6.46	5.90	6.27
1967	2.14	2.89	2.39	1979	7.23	6.72	7.06
1968	2.80	3.52	3.04	1980	7.97	8.81	8.25
1969	3.38	4.32	3.69	1981	8.11	10.78	8.99
1970	3.86	5.26	4.33	1982	7.43	10.45	8.43
1971	4.05	4.66	4.25	1983	6.23	8.90	7.11
1972	3.95	4.50	4.13	1984	2.40	9.50	4.74

Column 1: Weighted moving average of past inflation rates.
Column 2: AAA bond rate less 2.5 percentage points.
Column 3: Weighted average of (1) and (2) with weights of 2/3 and 1/3, respectively.

sum of column 1 weighted by two-thirds and column 2 weighted by one-third produces reasonable values for the expected rate of inflation in each of the years listed.

APPENDIX 12D: DERIVATION OF COST OF EQUITY

According to our definition, the marginal cost of equity after taxes can be obtained from the expression

$$MVE_0 = \int_0^{\infty} D_t e^{-\rho t}\, dt \qquad (12D.1)$$

where MVE_0 is the market value of the company's equity at time $t = 0$, D_t the dividend value expected to be paid at time t, and ρ the nominal cost of equity after corporate taxes.

To solve Eq. 12D.1 for ρ, we must predict the dividend rate from the beginning to the distant future. Dividends in successively more distant years are increasingly more difficult to predict, but fortunately they have a diminishing effect on the results, particularly for relatively high values of the cost of equity ρ.

A simple solution for ρ can be obtained if we assume a constant rate of growth, g, for future dividends so that

$$D = D_0 e^{gt}. \tag{12D.2}$$

Accordingly, Eq. 12D.1 becomes

$$MVE_0 = D_0 \int_0^\infty e^{(g-\rho)t} dt \tag{12D.3}$$

and after integration and rearrangement

$$\rho - g = \frac{D_0}{MVE_0}. \tag{12D.4}$$

The ratio $\frac{D_0}{MVE_0}$ is the initial dividend yield y. It follows that

$$\rho = y + g. \tag{12D.5}$$

In general, neither the earnings nor the dividends of a corporation or sector will grow at a constant rate indefinitely. Moreover, companies do not always pay a dividend that is proportional to earnings. For example, the optimum policy for a company growing much faster than the average may be to pay little or no dividends in its early stages in order to finance its rapid expansion, and then begin to pay a large dividend when growth subsides. Of course, this policy is beneficial if the company can reinvest its earnings at returns greater than the average cost of equity. It is clear that a specific model for dividend projections is needed to determine the cost of equity for any given company or sector.

A steady growth projection is a reasonable approximation when applied to a large group of companies, or to a large business sector because individual variations in dividend growth do not affect the average results for the group.[16] We have assumed a steady growth model (Appendix 12E) for the evaluation of the cost of equity to the aggregate of U.S. nonfinancial corporations.

During any particular year, we calculate an instantaneous yield, y, by dividing cyclically adjusted aggregate dividends paid by the sector during the year by the arithmetic average of the market value of the equity at the beginning of the year and that at the end of the year augmented by the dividends paid.

In Eq. 12D.5, the projected growth, g, refers to the growth in dividends stemming from the equity existing during the year in question and not the growth resulting from issuance of new equity. We call it the sustainable growth rate, and discuss the procedure for its evaluation in Appendix 12E.

For growth companies, many of which pay negligible dividends, procedures for calculating the cost of equity can be very complicated. One such procedure involves estimation of the growth rate and the length of the period during which the company is expected to grow faster than the corporate sector as a whole. It is assumed that after that period, the company reaches a mature state where its future growth prospects appear the same as those of an average company.

Application of this procedure to a hypothetical company whose price-to-earnings ratio and expected eighteen-year growth rate are twice those of the corporate sector, yields a cost of equity about equal to that of the corporate sector. This example illustrates the fact that if a company has a relatively high price-to-earnings ratio it does not necessarily mean that its cost of equity is relatively small. It also supplies a theoretical rationale for the rule of thumb used by many stock analysts to the effect that a fair price-to-earnings ratio for a company is about equal to its projected growth rate expressed in percent.

Data used in calculating costs of equity in the United States and Japan are summarized in Tables 12D−1 and 12D−2, respectively. The growth rates and the dividend yields for both countries were taken from the entire corporate sector, not just principal manufacturers (see Appendix 12E). This reflects the assumption that over the longer term the equity cost of these manufacturing firms must be the same as for all corporations.

The U.S. dividend yield and returns were both smoothed using a simple procedure that regressed each concept on the gap between potential and actual gross national product, and on a time trend. The smoothed values for each series were then taken as the estimated values from their equations with the gap set equal to its postwar average. This procedure is identical to that described in detail by Hatsopoulos (1983).

There is a further adjustment required in calculating the cost of equity in Japan. Since in Japan part of the dividends paid are tax deductible to a corporation (see Appendix 12B), the after-tax cost to the corporation is reduced by $\alpha\tau D_0$, where α is the fraction of the

Table 12D–1. Values of Parameters Used in Cost-of-Equity Calculations for the United States, (%).

Year	Projected Sustainable Real Growth Rate	Projected Inflation Rate	Dividend Yield	Cost of Equity After Taxes
1961	2.16	1.75	3.39	7.30
1962	2.52	1.68	3.65	7.85
1963	2.54	1.59	3.36	7.50
1964	2.96	1.58	3.01	7.55
1965	3.17	1.65	2.96	7.78
1966	2.48	2.15	3.27	7.90
1967	2.92	2.39	3.05	8.36
1968	2.81	3.04	2.84	8.68
1969	2.06	3.69	3.05	8.81
1970	1.81	4.33	3.80	9.93
1971	2.38	4.25	3.33	9.95
1972	2.33	4.13	3.17	9.63
1973	2.35	4.60	3.66	10.60
1974	2.43	6.13	5.67	14.22
1975	2.52	6.50	5.99	15.01
1976	2.55	6.06	5.28	13.89
1977	2.61	5.84	5.68	14.13
1978	2.58	6.27	6.24	15.09
1979	2.75	7.06	6.45	16.27
1980	2.50	8.25	5.80	16.55
1981	3.14	8.99	5.49	17.62
1982	2.96	8.43	6.16	17.55
1983	3.11	7.11	5.50	15.72
1984	2.31	4.74	4.82	11.88

dividend that is deductible and τ the tax rate at the margin. It follows that the term $\frac{D_0}{MVE_0}$ encountered in Eq. 12D.4 equals $(1 - \alpha\tau)$ times the yield y and, therefore, the applicable equation for the cost of equity in Japan is

$$\rho = (1 - \alpha\tau)y + g. \qquad (12D.6)$$

Table 12D–2. Values of Parameters Used in Cost-of-Equity Calculations for Japan, (%).

Year	Projected Sustainable Real Growth Rate	Projected Inflation Rate	Dividend Yield[a]	Cost of Equity After Taxes
1961	4.00	6.41	4.44	14.85
1962	4.00	6.36	4.44	14.80
1963	4.00	5.60	4.44	14.04
1964	4.00	5.80	4.44	14.24
1965	4.00	5.45	4.44	13.89
1966	4.00	5.31	4.24	13.55
1967	4.00	5.21	3.94	13.14
1968	4.00	5.38	3.62	13.00
1969	4.00	5.32	3.32	12.64
1970	4.00	5.97	2.98	12.95
1971	4.00	5.69	2.74	12.44
1972	4.00	5.17	2.54	11.71
1973	4.00	6.75	2.33	13.08
1974	4.00	10.69	2.14	16.82
1975	4.00	10.08	1.90	15.99
1976	4.00	9.29	1.69	14.98
1977	4.00	8.04	1.47	13.50
1978	4.00	6.81	1.36	12.17
1979	4.00	6.28	1.32	11.59
1980	4.00	5.88	1.31	11.19
1981	4.00	5.39	1.30	10.69
1982	4.00	4.72	1.29	10.01
1983	4.00	3.99	1.24	9.23
1984	4.00	3.66	1.17	8.83

a. Adjusted for "dividend deductibility" (see text).

APPENDIX 12E: SUSTAINABLE GROWTH IN DIVIDENDS PER SHARE

The sustainable growth in dividends per share is an important determinant of the cost of equity. One technique for estimating growth would be to inspect the historical growth in dividends per share and simply assume that investors expected that the recent growth would continue indefinitely. This approach has two primary faults: first, it is retrospective not prospective; second, it ignores the effects of tax rules, investment decisions, interest rates, expected inflation, borrowing capacity, etc. on the sustainable growth in per-share dividends. An alternative approach was used which directly estimates the growth rate of dividends that can be sustained by a corporation in the steady state by means of a combination of growth in retained earnings and by proportional increase in the amount of debt per share.

Under steady-state conditions, all entries in the balance sheet, profit-and-loss statement, and cash flow will remain in the same proportions. This implies that the rates of return, interest rates, inflation rates, tax schedules, dividend pay-out ratios, and other financial ratios are expected to remain constant in subsequent years. The constant growth rate of dividends that is attributable to retained earnings plus a proportional issuance of debt will be called the *sustainable growth rate*.

In the steady state, the growth rate of dividends is the same as that of net capital employed. Therefore, its relation to other financial parameters can be derived by considering the time evolution of the net capital employed. The change in net capital employed from year to year is equal to gross profits, minus capital consumption, minus interest, minus taxes, minus dividends, plus change in debt. Each of these elements can be expressed as a function of the average net capital employed. For instance, gross profits is derived from the gross return on capital employed. Capital consumption is derived directly from the share of fixed depreciable assets in the total capital employed. Interest costs are derived by multiplying interest rates by the product of the debt-to-capital ratio and capital employed, and so on.

The actual derivation of the sustainable growth is a complex simultaneous calculation. It has been explained at length in Hatsopoulos (1983).

Growth calculations for the U.S. nonfinancial sector produced very reasonable results that were roughly consistent with the actual growth in that sector over the period studied. Unfortunately, calculations of sustainable growth for the Japanese corporate sector produced results which were not consistent either with the historical record or with reasonable estimates of per-share growth. A number of factors can account for the failure of these growth calculations for Japanese corporations. First, there has not been the same stability in the financial ratios of Japanese corporations as that seen in U.S. corporations. Because of this, the growth calculations produce inconsistent results. Any change in financial ratios will alter the growth prospects in ways that the sustained growth calculation will be unable to pick up. Second, much more of the Japanese corporate sector's growth resulted from the injection of outside resources, either through direct subsidies or much greater "debt" financing by the primary equity holders. This tends to distort the results of the growth calculation which is designed to estimate potential growth on a per share basis.

After inspecting relevant trends in the value of corporate sales, income, and dividends per share, a reasonable growth assumption of 4.0 percent per year was adopted for the Japanese corporate sector.

APPENDIX 12F: COST OF INTANGIBLE CAPITAL

The general concept used to calculate the cost of capital for each type of asset was described in the text of the chapter. The approach used is an extension of that used by Hall and Jorgenson (1967).

The Hall and Jorgenson concept requires that the cash paid to acquire an asset, the tax credit associated with the holding of the asset, and the maintenance cost of the net cash employed be distributed over time in proportion to the economic use of the asset over time. In order to arrive at an appropriate expression for the cost of capital services, one needs to define a time profile for the "degree of usefulness" of the asset after acquisition. In their paper, Hall and Jorgenson assumed a profile that decays at a constant rate as illustrated in Figure 12F−1a. Their concept (but not their equation) applies equally to an asset, such as a light bulb, whose usefulness profile is best represented by Figure 12F−1b.

Investments in knowledge invariably involve a delay between the time of acquisition of the knowledge and the time of its application. The appropriate profile for such investments is shown in Figure 12F−1c. In drawing this curve, it was assumed that the usefulness of a given knowledge decays with time as it is superseded by other developments. In principle, time delays are also encountered in tangible investment such as the construction of a chemical plant. They are, however, usually short enough to be neglected. A notable exception would be the construction of a nuclear power plant. In such a case, as well as in long-term investments in knowledge, time delays can have a significant effect on the cost of capital services, and must be taken into account.

For each investment there is an appropriate profile for the cash flow resulting from tax benefits. For knowledge investments, tax benefits accrue immediately upon acquisition, because R&D costs are deductible

Figure 12F−1. Time Profile of Degree of Usefulness of an Asset (F = Degree of Usefulness in Percent; t = Time after Acquisition)

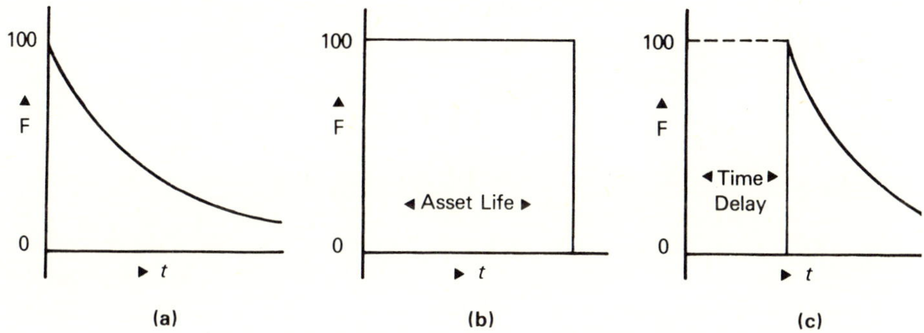

in the year they are incurred. For tangible investments, on the other hand, tax benefits accrue over time as specified by the law.

The concept of a "cost of knowledge capital" is applicable, in principle, not only to R&D investments but also to other intangible investments such as market development.

Two other modifications of the Hall and Jorgenson analysis were implemented in the present work. The first was a modification to account for inflation, and the second a modification to reflect the timing of tax payments and of tax credits.

The actual equation is very complicated. However, utilizing several simplifying assumptions and ignoring the timing of taxes, the equation reduces to

$$c = (1 + \rho - \pi)^d \; \frac{[1 - ITC - \tau(1 - \beta)z](\rho + \delta - \pi)}{(1 - \tau)} \qquad (12E.1)$$

where:

- c = the cost of capital
- ρ = the cost of funds
- π = the expected inflation rate
- d = the time delay before the investment may be applied
- ITC = the Investment Tax Credit
- τ = the corporate tax rate
- β = the portion of the ITC that reduces the basis for tax depreciation
- z = the present value of depreciation allowances
- δ = the rate of economic decay

APPENDIX 12G: DECAY RATES OF ASSETS

For the United States, instantaneous average decay rates (economic depreciation) for equipment and structures in the manufacturing sector are listed in Table 12G–1. The data are derived from the U.S. Department of Commerce (1982a, b). They are calculated by dividing the annual total for depreciation by the average of the current and prior year's end-of-year net stocks of each asset. In our calculations we have assumed a fixed rate of 13.1 percent for equipment and 6.8 percent for structures.

In calculating the cost of capital in Japan, these same decay rates were used despite the fact that the mix of Japanese fixed assets is different from that in the United States. This was done in order to estimate the cost of capital that would apply to U.S. firms (and the capital they purchase) if they had been operating in the same financial environment and with the same tax rates as Japanese firms.

Details on tax rates and depreciation schedules in Japan can be found in Appendix 12B, especially Table 12B–2.

Table 12G–1. Instantaneous Depreciation Rates in the Manufacturing Sector (% per Year).

Year	Equipment	Structures	Year	Equipment	Structures
1960	13.50	6.51	1973	12.83	6.65
1961	13.63	6.53	1974	12.38	6.79
1962	13.73	6.59	1975	12.64	6.86
1963	13.77	6.67	1976	12.53	6.81
1964	13.75	6.69	1977	12.61	6.79
1965	13.58	6.69	1978	12.52	6.80
1966	13.35	6.64	1979	12.63	6.91
1967	13.15	6.58	1980	12.65	6.96
1968	13.06	6.54	1981	12.83	6.88
1969	12.92	6.58	1982	13.07	6.84
1970	12.92	6.56	1983	13.35	6.87
1971	12.96	6.66	1984	13.67	7.05
1972	12.96	6.67			

NOTES

1. We estimate that U.S. manufacturers invest about the same amount of dollars in tangible assets as in intangibles.
2. The average standard of living of any nation is strictly dependent on the rate of growth of its productivity. Productivity in manufacturing is twice as high as productivity in the service sector. For this reason, growth of the manufacturing sector is critical to growth in productivity. For example, if productivity in each of manufacturing and services grew at 1 percent per year but employment in manufacturing stayed the same while employment in services increased by 10 percent, the overall productivity of the country would decline by 1 percent.
3. Earlier studies of the cost of capital (Hatsopoulos 1983, 1984) included net receivables as a use of funds. By contrast, this study assumes that net payables (the negative of net receivables) are a source of funds. To these are added other deferred payments

(mostly deferred taxes) to calculate a corporation's net interest-free debt. While it is not traditional to think of these as marginal sources of funds for corporations, an incremental expansion of the firm would certainly include some increases in these categories of deferred payments.

4. The Chase results are based on Dodwell Marketing Consultants (1978).
5. This fact is not always recognized by managers. See Auerbach (1984).
6. The earlier study of the cost of capital (Hatsopoulos 1983, 1984) assumed that debt was the exclusive marginal source of funds for Japanese corporations. This study relaxes that assumption and instead chooses a debt-equity mix that is consistent with the balance sheets of Japanese principal corporations. However, if a Japanese firm uses retained earnings to repay debt (because it cannot repurchase equity), then retained earnings should be "priced" as debt, not equity, and the debt-equity assumption used here will understate the use of debt and overstate the use of equity.
7. Data on inventory obsolescence are difficult to interpret. Case study evidence suggests a shrinkage and osbolescence cost of about 4 percent (see Hall 1974). However, our review of National Income and Product Accounts data suggested a somewhat smaller effect.
8. Many corporations pay taxes on inventory appreciation. However, if a firm is on LIFO (last in, first out) accounting and its inventories are growing steadily, it will pay no taxes on inventory appreciation. Accordingly, in evaluating marginal costs associated with growth, capital gains on inventories were assumed to be not taxable.
9. There are three factors that can account for the high price-to-earnings ratios: the tax treatment of capital gains, the projected (or expected) growth in the earnings of Japanese corporations, and the tax deductibility in Japan of a portion of dividends received by individuals (Shoven and Tachibanaki 1985).
10. The May 28, 1985 tax proposal of the Reagan administration (dubbed "Treasury II") addresses some of the objectives described in this chapter. By eliminating some of the existing tax preferences, it makes the cost of capital more neutral across different types of assets and reduces consumption subsidies. On the other hand, it is doubtful that the proposal will have a major effect on the overall cost of capital or the national saving rate. Nevertheless,

the proposal could be modified in a way that would reduce the marginal cost of capital and increase national saving with virtually no effect on Treasury revenues. The proposed deduction of 10 percent of corporate dividends should be "traded" for a provision that would (1) make dividends on qualifying new preferred equity issues deductible to the issuing corporation and (2) increase the capital gains exclusion on qualifying common stocks to 100 percent. Restrictions on the kinds of securities that would qualify for these benefits could limit both the substitution of new equity for existing equity and the escape from taxation through corporate liquidation.

11. The analysis from this source is consistent with estimates for the entire nonfinancial corporate sector from the Flow of Funds Statistics of the Federal Reserve Board.
12. This is the series calculated by Feldstein, Dicks-Mireaux, and Poterba (1983).
13. This is the tax rate in Tokyo. In other cities and small municipalities, the tax rate for inhabitants' tax may be lower.
14. This is the tax rate applicable for the late 1970s. While we believe it to reasonably represent tax rates in earlier periods, there might be minor changes.
15. Another estimate could be obtained by assuming that investors seek a real interest rate after taxes, i^*, in constant dollars. Then, the rate of expected inflation π^e would be $\pi^e = i(1 - \tau) - i^*$, where τ is the investors tax rate.
16. In general, some firms are in the high growth phase, others in the payout period, and still others are declining. Provided this mix is fairly constant, the average dividend will grow at a constant rate.

REFERENCES

Auerbach, A.J. 1984. "Taxes, Firm Financial Policy and the Cost of Capital: An Empirical Analysis." *Journal of Public Economics 23:* 27–57.

Blades, D.W. and P.H. Sturm. 1982. "The Concept and Measurement of Savings: The United States and Other Industrialized Countries." In Federal Reserve Bank, Boston, *Savings and Government Policy* (October): 1–45. Boston.

Chase Financial Policy. 1980. "U.S. and Japanese Semiconductor Industries: A Financial Comparison." NewYork: Chase Financial Policy, a Division of the Chase Manhattan Bank.

Denison, E.F. 1974. *Accounting for United States Economic Growth 1929–1969.* Washington, D.C.: The Brookings Institution.

Dodwell Marketing Consultants. 1978. *Industrial Groupings in Japan.* Rev. ed.

Elston, C.D. 1981. "The Financing of Japanese Indusry." *Bank of England Quarterly Bulletin* (December). London.

Feldstein, M., L. Dicks-Mireaux, and J. Poterba. 1983. "The Effective Tax Rate and the Pretax Rate of Return." *Journal of Public Economics 21:* 129–158.

Feldstein, M. and L. Summers. 1978. "Inflation, Tax Rules, and the Long Term Interest Rate." *Brookings Papers on Economic Activity 1:* 61–69.

Hall, R.E., and D.W. Jorgenson. 1967. "Tax Policy and Investment Behavior." *American Economic Review 57*, no. 3 (June): 391–414.

Hall, T. 1974. "Inventory Carrying Costs: A Case Study." *Management Accounting* (January): 37–39.

Hatsopoulos, G.N. 1983. "High Cost of Capital: Handicap of American Industry." American Business Conference and Thermo Electron Corporation, April 26, 1983.

——— 1984. Memorandum of January 17, 1984 to Presidential Commission on Industrial Competitiveness. (Updates Hatsopoulos 1983.)

Modigliani, F., and R.A. Cohn. 1979. "Inflation, Rational Valuation, and the Market." *Financial Analyst's Journal* (March/April).

National Income and Products Accounts. 1981. "1929–76 Statistical Tables." United States Department of Commerce, Bureau of Economic Analysis, Washington, D.C.

——— 1982. "Survey of Current Business." United States Department of Commerce, Bureau of Economic Analysis, Washington, D.C. Vol. 62, no. 7 (July).

——— 1984. "Survey of Current Business." United States Department of Commerce, Bureau of Economic Analysis, Washington, D.C. Vol. 64, no. 7 (July).

Organization for Economic Cooperation and Development. 1984. "National Accounts, 1970–1982." Paris: Organization for Economic Cooperation and Development.

Shoven, J.B., and T. Tachibanaki. 1985. "The Taxation of Income from Capital in Japan." Paper presented at the Economic Policy Research Conference on Government Policy Towards Industry in the United States and Japan, May 2–3.

Touche-Ross International. 1980. *Business Study: Japan.* New York.

U.S. Department of Commerce. 1982a. *Fixed Reproducible Tangible Wealth in the United States, 1925–1979.* Washington, D.C.: U.S. Department of Commerce, Bureau of Economic Analysis, March.

——— 1982b. *Fixed Reproducible Tangible Wealth in the United States, 1980–1981.* Washington, D.C.: U.S. Department of Commerce, Bureau of Economic Analysis, October.

13 THE RELATIVE IMPACT OF TAXATION AND INTEREST RATES ON THE COST OF CAPITAL

Aldona E. Robbins, Gary A. Robbins, and Paul Craig Roberts

The chapter calculates the relative impact of interest rates, taxes, and technology on the cost of capital. The results show that the cost of capital is highly inelastic with respect to changes in the interest rate. Taxation, however, substantially affects the cost of capital, raising it by 43 percent for the economy as a whole. Increases in the cost of capital due to tax changes raise the rate of return required from real productive assets, which translates into fewer viable investment opportunities.

The results of this chapter dispute the policy prescription implied by the view that federal deficits cause high interest rates, which in turn, crowd out investment. The view that higher taxes would reduce crowding out and raise the investment rate is inconsistent with our findings that taxation has significant adverse effects on the rate of capital formation.

* This chapter is part of ongoing research efforts. The authors would like to thank David Brazell for his contributions to the theoretical and empirical work in the paper; William Dahl for his assistance with the empirical presentation; and Stephen Entin, Manuel Johnson, Michael Kaufman, and Norman Ture for their insights and comments. The authors accept sole responsibility for any errors.

A SUPPLY-SIDE FRAMEWORK FOR ANALYZING PRODUCTION COSTS

Since the time of the New Deal, U.S. economic policy had been heavily influenced by Keynesian theory. Policymakers thought that by following demand management strategies of raising or lowering taxes and thereby increasing or decreasing the size of government, the economy could be kept on a stable growth path. Aggregate economic activity was believed to depend only upon *average* tax rates. Prices, other than the general price level, did not matter. In this framework only the level of disposable income for the economy as whole determines growth. Consequently, it was thought possible to redistribute income costlessly from one person or group to another by transferring income via the tax and welfare systems.

Problems ensued, however. Neither the business cycle nor poverty was eliminated. It appeared that a trade-off existed between unemployment and inflation, the so-called Phillips curve. Unemployment, it seemed, could only be driven lower at the cost of higher inflation. Then a most puzzling event occurred. During the 1970s the economy experienced what came to be known as stagflation—the *simultaneous* occurrence of economic stagnation, high unemployment, and high inflation.

These events readied the stage for the emergence of what became known as supply-side economics. Although its name was presumably to distinguish it from Keynesian demand-side policies, it has led to some misconceptions. In fact supply-side economics is really just the application of the *microeconomic* theory of consumption and production to the aggregate economy. Supply-side theory posits that what is observed at the macroeconomic level is really the sum of all the individual microeconomic transactions in the economy. Thus, the explanation of overall economic behavior actually lies in the myriad of markets which it encompasses.

According to the principles of microeconomics, markets are composed of buyers with demand schedules and sellers with supply schedules. These schedules are merely lists of quantities of various goods or services that buyers are willing to purchase and sellers are willing to sell at various prices. For each good there will be one price at which the buyers' and sellers' quantities match, and the transaction is made.

The relevant point here, however, is that the buyers and sellers are responding to prices. Behind the derivation of each of these schedules is some very elegant theory which takes into account, among other things, the prices of all other goods and services. If any of these *relative prices* changes, the demand and supply schedules also change, unlike the textbook Keynesian framework where relative prices do not matter.

One of the basic premises of supply-side economics is that government policy affects the relative prices of factors of production: land, labor, and capital. During the 1970s, the effect of government policy was to raise the cost of capital and lower the return to labor by increasing the tax burden at all levels of government. The expansion of the government sector relative to the private sector thereby reduced the incentive to invest and to work.

SUPPLY-SIDE ECONOMICS AND THE COST OF LABOR

Although the focus of this chapter is on capital, labor is a factor of production also affected by taxes. The impact of taxes on labor costs is easier to analyze because labor compensation generally occurs at the same time labor services are provided.

Workers generally measure their well-being in after-tax terms, namely the gross wage rate less income and payroll taxes. Any increase in income or payroll taxes reduces the worker's after-tax wage rate and well-being unless the rate is raised by an appropriate amount. If this does not happen, an increase in tax rates causes the supply schedule of labor to shift up and to the left. The general consensus of econometric studies is that a 10 percent reduction in take-home pay reduces labor supply by about 4 percent.

Labor costs to the firm increase directly by any increase in employer payroll taxes and by any increase in wage rate that occurs as workers pass a portion of their income or payroll tax increases on to the firms in order to keep their after-tax wage from falling. As the wage rate rises, labor becomes more expensive relative to capital, and firms will substitute capital for labor where possible, resulting in lower employment.

The disparity between the gross before-tax rate of labor compensation and the after-tax wage of workers widened during the 1970s because of increases in income and payroll tax rates. The average marginal tax rate on U.S. wages rose from 21.2 percent in 1965 to 38.3 percent in 1981, almost a doubling. The increase due to income taxes alone was from 17.8 percent to 27.5 percent. Much of this rise was due to the silent tax increase known as "bracket creep." Because the U.S. federal income tax system is progressive, additions to income are taxed at higher and higher rates. As workers' wages rise with inflation, they are pushed into higher tax brackets even though their real income has not changed. Consequently, the rapid inflation of the 1970s meant that firms had to increase the gross wage rate *faster* than the rate of inflation simply to keep the worker in the same after-tax position.

Payroll taxes were growing even faster than income taxes. In the United States, payroll taxes are used primarily to finance the Social Security system. Social Security required higher taxes for sufficient financing because real levels of cash benefits were increased, Medicare was added, and the pay-as-you-go system matured. In 1965 the combined employer-employee tax rate was 7.25 percent on the first $4,800 in wages. By 1981 this tax rate had risen to 13.3 percent on the first $29,700 in wages. Not only did the tax rate almost double, but the share of wages in the United States subject to the tax rose from 72 percent to 90 percent because of the sharp increase in the wage base. As in the case of bracket creep, the firm had to continually increase the worker's gross wage rate in order to keep the worker's after-payroll-tax wage the same.

The effect of rising average marginal income tax rates and payroll tax rates caused the cost of labor to increase without any compensating increases in productivity. As the price of labor rises relative to capital, microeconomic theory predicts that the firm will substitute capital for labor where possible or reduce output. Capital, however, was subject to the same federal income tax system. The remainder of the paper looks at what was happening to the cost of capital.

FINANCIAL INSTRUMENTS AND THE INVESTMENT DECISION

Central to the topic of technology and economic policy is the impact of forces external to the economic decisionmaker, including government policy, on the investment process. It is impossible to discuss the factors affecting economic progress without placing them in the context of their influence on the basic determinants of behavior. "Shocks" must be gauged in terms of their influence on the supply and demand for goods and services in order to gain any insight into their final effect.

Impacts on the supply of labor are the most easily seen since the provision of and payment for labor services are more or less coincident. For example, a shock which affects the labor contract can be judged in terms of the likely effect on the labor supply schedule and can be quickly analyzed through the easily observable impacts on wages and employment.

The analysis of the impact of shocks on the supply of capital services is much less straightforward because its pattern of compensation spans more than one period. The analyst must develop a method to translate a multiperiod investment compensation pattern into current period equivalents. This falls under the general rubric of the service price. The analysis will now proceed by examining the financial investment decision in detail. A set of principles will be developed that will ultimately be applied to the purchase of a real capital asset.

Consol with Perfect Foresight

Zero Inflation, No Tax. First the textbook example of a *consol*—an asset which yields an infinitely lived, constant stream of earnings—is examined. At the margin the net payment per period from the consol divided by its purchase price equals the discount rate for the individual as well as for the market. Knowing two of the three elements of the transaction uniquely determines the third. For example, in order to induce an investor with a discount rate of 4 percent to purchase a consol of $1,000, the net payment, or debt service, per period must be $40. The formula for the debt service in the case of a simple consol is written as

$$\text{Consol debt service} = \$1 \times r, \tag{13.1}$$

where r is the discount rate.

The discount rate is the rate of exchange between current and future consumption for the individual and for the market. To avoid unnecessary complications, the remaining analysis assumes that the discount rate remains constant through time. The formulas will employ continuous compounding, and the purchase price of the assets portrayed will be $1.00.

Inflation, No Tax. The presence of inflation means that the value of the goods that can be purchased with the proceeds from the consol changes from period to period. Future net payments must, therefore, be adjusted for changes in purchasing power. If inflation is expected to run at 5 percent (forever), the value of all future payments and therefore the principal from any subsequent sale is reduced by 5 percent each year, and the debt service payment in the previous example would have to increase from \$40 per period to \$90 to keep the investor no worse off after inflation. Thus, the relation between the debt service payment and the purchase price is modified to be the discount rate *plus* the inflation rate. This is normally termed an "inflation premium." The debt service in this case is written as

$$\text{Consol with inflation debt service} = \$1.00 \times (r + z), \quad (13.2)$$

where z is the expected inflation rate.

Inflation and a Tax. The presence of a tax directly reduces the net payment received by the investor, requiring a higher pre-tax payment to keep the investor as well off as before. Assuming a 50 percent tax rate, half of the proceeds of each payment goes to the taxing authority. From the preceding example the payment amount would have to double from \$90 to \$180 per period. The debt service formula is modified to be

$$\text{Consol with inflation and tax debt service} = \$1.00 \times (r + z)/(1 - t), \quad (13.3)$$

where t is the expected marginal tax rate.

Now the mathematical relation becomes the discount rate plus the inflation rate *all divided by* 1 minus the tax rate to translate the after-tax amounts to necessary pre-tax amounts. This translation is used in the remainder of the chapter to indicate the augmentation of after-tax to pre-tax amounts. *Note that the tax is levied on the inflation premium as well as the pure interest element of the consol.* This phenomenon, referred to as the Fisher effect, explains an important interaction between taxes, inflation, and the interest rate. As either taxes or inflation increase, the nominal interest rate must increase at a faster pace than the simple sum of the previous two rates.

Purchase of a Bond with Perfect Foresight

The previous analysis can be easily extended from an infinite payment stream to one with a fixed time horizon and repayment of principal. In this case, the previous examples can be thought of as the equivalent of two simple transactions: (1) purchase of the consol and (2) subsequent sale at a price agreed to at the time of purchase. Under conditions of constant inflation and rate of time preference, the geometry of the bond analysis and the consol are identical. Varying either the inflation rate or discount rate will complicate the problem but not materially change the conclusions. The resulting formula for a bond's debt service is

$$\begin{aligned}\text{Bond debt Service} &= \$1.00 \times (1 - e^{-(r+z)\times T}) \\ &\times ((r + z)/(1 - e^{-(r+z)\times T})/(1 - t) \\ &= \$1.00 \times (r + z)/(1 - t),\end{aligned} \tag{13.4}$$

where T is the maturity date.

The payment amount necessary to attract the investor in our example is the original purchase price of the instrument less the inflation-adjusted, discounted value of the future return of principal. The term *net-of-tax private present costs* is used to denote the net present value of costs and repayments that are not directly related to the income stream of the asset—in this case, the original purchase price less the present value of the return of principal. This allows the analysis to concentrate separately on the net acquisition/principal repayment and the periodic income, the two major aspects of the transaction.

This valuation of the net-of-tax private present cost of the bond must be recovered over the term of the bond along with the net interest income required by the investor. The present value of the payment stream will be used to characterize both the level and pattern of the periodic income stream. The general formula for the debt service of an investment is the net-of-tax private present cost of the investment, $(1 - e^{-(r+z)\times T})$, translated to pre-tax levels and divided by the present value of the payment stream, $(1 - e^{-1(r+z)\times T})/(r + z)$. This can be seen to reduce to our previous example, and the debt service will be exactly as before, or $180 per period.

Empirical data support the behavior of nominal interest rates, inflation, and taxes just posited (Table 13–1). Figure 13–1 plots the nominal U.S. twenty-year Treasury bond rate against the rate of increase of the private gross national product deflator over the period

Table 13–1. Data for Figure 13–1.

Year	Private GNP Deflator (%)	20–Year Treasury Rate (%)
1954	1.53	2.64
1955	2.11	2.90
1956	3.73	3.14
1957	2.64	3.54
1958	1.10	3.48
1959	2.28	4.13
1960	1.09	4.06
1961	0.71	3.92
1962	1.74	3.99
1963	1.11	4.05
1964	1.04	4.19
1965	1.99	4.27
1966	3.75	4.77
1967	2.54	5.01
1968	4.42	5.45
1969	5.22	6.33
1970	4.32	6.86
1971	4.08	6.12
1972	3.45	6.01
1973	7.10	7.12
1974	10.77	8.05
1975	7.51	8.19
1976	4.35	7.86
1977	5.95	7.67
1978	8.80	8.48
1979	8.36	9.33
1980	10.19	11.39
1981	8.74	13.72
1982	3.92	12.92
1983	3.84	11.34

1954–1983. As expected, nominal interest rates and inflation move in somewhat the same direction. Some analysts take the difference of these two series and call it the *real interest rate*. Two noteworthy observations about this graph should be made: (1) there are several periods when the real interest rate is negative, and (2) taxes have been ignored.

Figure 13–1. Nominal Interest Rates and Inflation (1954–1983).

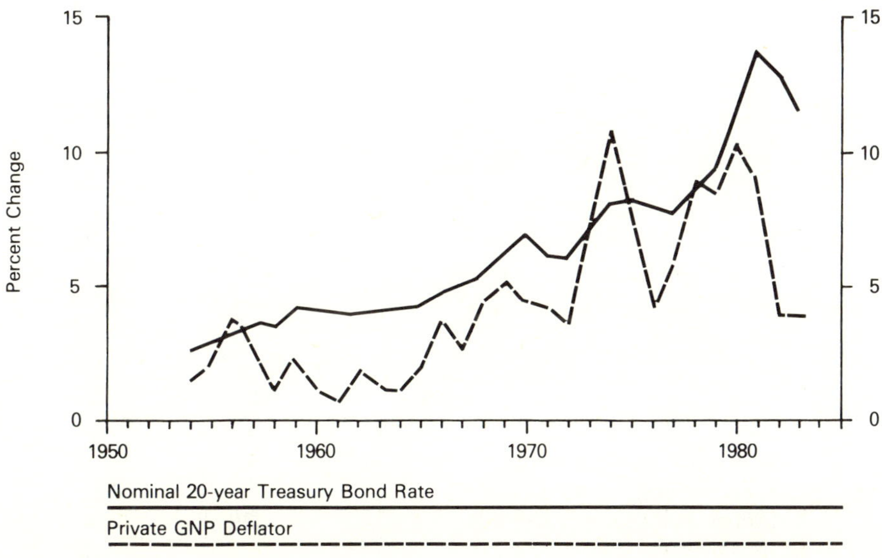

Figure 13–2 adjusts the nominal interest rate for taxes using the weighted average marginal tax rate on interest income at the federal, state, and local levels prevailing in the United States at each period and plots it against inflation. The difference between these two series is referred to as the *real after-tax interest rate.* after-tax nominal interest rates follow the rate of inflation more closely than the before-tax rates, and there are even longer periods of negative returns to ownership of U.S. Treasury bonds. In order to explain these periods we must look to other causes of this divergence.

Purchase of Bond with Risk

Up to this point in our analysis we have assumed that the investor knows all current and future prices. However, this is not the way the real world operates. Rather, the investor and the market must make

Table 13–2. Service Prices for U.S. Corporate Nonresidential Depreciable Investment under Current Law, 1984 ACRS (%).

Service Price per Unit of Output

	Equipment	Structures	Total
Technology only	18.99	6.04	14.61
Property tax only	22.21	9.26	17.83
Sales tax only	20.03	6.37	15.41
Property and sales tax only	23.42	9.77	18.80
Income taxes only	20.64	8.60	16.57
All taxes	25.16	12.46	20.86

Relative Contribution of Technological and Tax Factors

	Equipment	Structures	Total
Technology only	75.49	48.49	70.03
Property tax only	12.80	25.83	15.43
Sales tax only	4.11	2.64	3.81
Property and sales taxes only	17.60	29.88	20.08
Income taxes only	6.55	20.52	9.38
Effective tax rates	24.51	51.51	29.97

guesses about future prices and taxes. These guesses are incorporated into our analysis as "expected" inflation rates and "expected" tax rates.

Expectations are generally characterized by a mean and by a standard deviation, or a measure of the likely dispersion of anticipated future prices. In the case of perfect knowledge, all the investor needs to know is the average inflation and tax rates, because their standard deviations are zero. In our current circumstance the investor must include a risk premium to cover the costs of making a wrong guess. This risk premium is directly related to the perceived standard deviation of future prices. Because a guess involves judgment on the part of an investor or the market, there is no guarantee that the guess will be right. Thus the presence of risk and of future inflation expectations differing from those currently prevailing can explain periods—even long ones—of apparent negative returns.

Figure 13–2. Nominal After-Tax Interest Rates and Inflation (1954–1983).

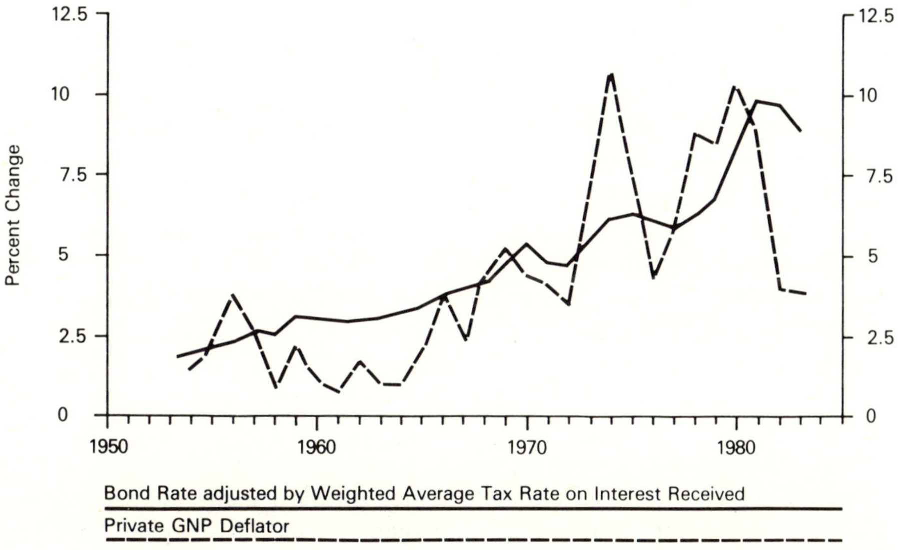

The relation has to be elaborated one step further to include a term reflecting the risk of misguessing inflation, the *risk premium.* This taxable risk premium must be added to the discount rate and inflation term. Assume that the risk premium is two percentage points. The value of the translation factor increases from 18 percent to 22 percent after tax. The debt service stream increases accordingly from \$180 to \$220. *Note that the Fisher effect directly extends to risk premiums as well as to inflation rates,* and the debt service formula becomes

$$\text{Bond with risk debt service} = \$1.00 \times (r + z + \text{risk})/(1 - t), \qquad (13.5)$$

where risk is the expected loss or risk.

Choice between Bonds with Differing Risk

The assumption that investors will always choose those investments they believe will yield the greatest net return extends the analytic framework to a wider application. Since the discount rate and price forecast must be identical for each investment alternative considered, the investor will always choose the one which yields the higher net return. At equilibrium, the after-tax net of risk rates of return must be equal between every paired alternative and therefore for all investments. If the returns are not equal, an arbitrage situation will exist which will be exploited by "selling" the asset with the lower return and "buying" the one with the higher return until the two expected returns are identical. Thus the market equilibrium condition requires that each asset compete with every other.

Arbitrage in a rational market means that the relevant market for the determination of prices or rates of return for financial instruments must include the entire set of instruments available or known to the population of investors. The degree of interdependence necessary to assure this result is surprisingly small, merely that at least one investor know that the opportunity exists. This is because one investor can buy and sell enough of the two disparate issues to drive their prices into line. Those who argue a persistent divergence must also argue that they themselves are irrational since they forgo the opportunity to make certain and instant profits from the knowledge they profess to have. It seems more realistic to proceed on the premise that the market will yield equal expected risk-adjusted after-tax returns for all the paired choices available. New issues must compete with other new issues as well as all preexisting issues. The relevant market for rate determination is the entire portfolio of choices available to investors, and the market supply, therefore, is the *stock* of assets not simply the flow of new issues.

International Choice: the Exchange Rate

The extension of this analysis to include investment across borders requires only that the price expectation process be elaborated to include the movements of more than one currency. As previously found, the investor must make some judgment about the value of future financial flows in order to choose between alternative investments. The denomination of the terms of the instrument may dictate that two translations be made before the transaction can be evaluated in terms of real goods or services. The exchange rate is a simple shorthand characterization of this dual forecasting problem. As in the previous example, investors must convert all future payments into their current units of account. This requires both a forecast of the likely future prices in their domestic currency and the likely exchange rate between their currency and that in which the instrument is paid. As before, these forecasts must be made in the face of uncertainty and therefore must include and provide explicit accounting for the likelihood of loss through the inclusion of risk premiums,

$$\text{Foreign bond debt service} = \$1.00 \times (r + z + \text{risk} + \text{diff})/(1 - t), \tag{13.6}$$

where diff is the expected differential inflation rate.

Because the typical investor does not possess the necessary information to make the exchange rate forecast, this is normally left to specialists who do little else. This fact does not diminish the degree of competition between instruments of different countries since competition only requires one person to arbitrage the market to a point where all issues compete with one another. Thus, in the final analysis, U.S. government issues compete with those of a major corporation. Although this competition requirement could be relaxed somewhat to allow small deviations to exist, it would add nothing to the analysis of market tendencies except needless complication of the mathematics.

REAL ASSETS AND THE INVESTMENT DECISION: DERIVATION OF A CAPITAL SERVICE PRICE

The analysis necessarily becomes more complicated when one moves to a discussion of investment in real assets. This is due to the problem of characterizing the productive life in the value of a real asset. Complications also occur because of the more intricate tax treatment of capital assets and their returns. These complexities include accounting for property taxes, indirect business taxes, multiple levels of direct taxation (both the corporate and personal levels, for example), investment credits, and capital cost recovery systems.

For a given real asset, one can calculate an implied rental rate, analogous to the coupon rate, which equates the rate of compensation that would accrue to the owner if the asset's productive services were to be sold in a competitive rental market. These rentals would have to be at levels sufficient to cover both the anticipated taxes and expected decline in the asset's productive capability while maintaining a "normal" (risk-inclusive) rate of return. This rate of return, net of risk, is the same return required in the prior examples. For a given tax regime and a known pattern of productive efficiency, these rentals may be summarized by a capital service price. This service price represents the minimum current *marginal value product* that must be earned by an additional dollar's investment in the asset in order for that extra investment to be undertaken.

The usual derivation of the service price assumes that investors maximize their prospective wealth position. Investment in each alternative asset continues until further increments no longer yield an increase to expected wealth. This method yields an expression for the service price that relates the necessary before-tax return to the required after-tax return just as in the prior financial examples. As mentioned, the service price must cover expected loss in the asset's value and yield the same net return as alternative investments.

Before describing the derivation of the service price, an important point regarding the use of debt financing must be addressed. Recall that the investor must measure the cost of the marginal dollar's worth of investment. That dollar may be obtained either through borrowing or through equity financing. However, there can be only one single cost of capital at equilibrium given the arbitrage requirement. If there were more than one, the investor would always choose the least-cost method of financing. But because both means of financing are observed, the cost of debt and equity financing must be equal at the margin. Thus, the service price calculations described below are equally valid for a leveraged as well as an equity-financed investment.

Constantly Decaying Asset with Perfect Foresight

As in the case of the financial assets, a simple investment can be used to introduce a framework for further analysis. An asset with a geometrically declining pattern of output is analogous to the simple consol. In this case the investor must receive a service flow sufficient to cover his rate of time preference plus the decline in value of the asset each period. The formula for the service price in this simple case is

$$\text{Capital service price} = \$1.00 \times (r + d) \tag{13.7}$$

where d is the percentage decline in the value of the asset in terms of future output.

Inflation and a Sales Tax. Since the service price is in terms of units of real output, there is no change needed in the formula to incorporate inflation. The presence of a sales tax, however, reduces the net payment received by the investor. The service price formula in (13.7) must be translated to pre-tax levels and becomes

$$\text{Capital service price} = \$1.00 \times (r + d)/(1 - ts) \tag{13.8}$$

where ts is the sales tax rate.

Risk can be incorporated at this point by defining r so that it contains both the discount rate and a risk premium.

Inflation and an Income Tax. An income tax introduces yet another term, *tax depreciation,* to the analysis. Unlike a financial asset, the principal of the investment in a real asset is not returned, and an adjustment in the income flow must be made to attempt to incorporate the change in net worth for income tax purposes. The U.S. income tax laws measure income for tax purposes by subtracting an arbitrary "depreciation deduction" from the gross returns to capital which reduces the net-of-tax private present cost of the investment. Tax depreciation, in contrast with the decline in the real productivity of the asset, is a *financial* asset in that it is a financial allowance against taxes over some specified time. Thus, the depreciation term must be valued with an inflation term and within the context of the tax savings it yields the investor. The capital service price of straight-line depreciation becomes

$$\text{Capital service price} = \$1.00 \times [\frac{(1-t)\times(1-e^{-(r+z)\times T})}{(r+z)}]\times[\frac{(r+d)}{(1-t)}], \tag{13.9a}$$

where T is the arbitrary tax life and t is the income tax rate.

Using names and symbols to simplify, Eq. (13.9a) may be rewritten as

$$\text{Capital service price} = \$1.00 \times (1-D)\times[\text{SP}/(1-t)] \qquad (13.9b)$$

where D is the present value of tax savings due to depreciation deductions and SP is the service price of the asset disregarding taxes from Eq. (13.7).

It must be remembered that each of these two shorthand terms represent more complex expressions containing r, the rate of time preference plus risk, and z, the expected inflation rate. Finally, each of the prior equations could be solved for r in terms of the existing service price to yield the *real after-tax rate of return* to capital, which is nothing more than the discount rate plus the risk of owning a real capital investment. Thus the analysis which follows will allow the description of technological factors and tax depreciation schedules to become extremely complex without obscuring the basic relation between the service price and the real rate of return to capital.

Inflation, Income, and Indirect Taxes. The complete service price calculation used in this study requires several steps to incorporate all the major features of taxation. In order to simplify the analysis and emphasize the relative importance of the various taxes applied to the returns to real capital, the influence of income taxes is considered first in isolation, and then indirect taxes, namely sales and property, are added.

The return on capital subject only to income tax can be expressed by subtracting the investment tax credit and the tax value of depreciation terms described previously from the gross costs of the asset,

$$\text{Capital service price} = \$1.00 \times (1 - \text{ITC} - D) \times \frac{\text{SP}}{(1-t)} \qquad (13.10a)$$

where ITC is the investment tax credit rate, D is the present value of tax depreciation, SP is the service price of the asset disregarding taxes, and t is the income tax rate.

In the United States, indirect taxes (property and sales) also add to the "tax wedge" between pre-tax market returns and after-tax returns to investors in real assets. It is important to realize that the returns to capital must be sufficient to pay sales or value-added taxes and property taxes. To account for these costs, the rate of property taxation must be added to the service price expression described above. Then, the entire expression must be translated to reflect taxes on the sales or production of final output. The entire formula for the service price is

$$\text{Capital service price} = \$1.00 \times [tp + (1 - \text{ITC} - td) \times \frac{\text{SP}}{(1 - t)}]/(1 - ts), \quad (13.10b)$$

where tp is property tax rate and ts is the sales tax rate.

Because property taxes at any point in time are levied on the after-tax discounted present value of the remaining productive stream, it is as if the taxing authority adds a fixed percentage to the private cost of acquiring the asset. The sales tax takes a fixed percentage from the flow which services the investment. Thus, property taxes add directly to the cost of acquiring the asset, and the sales tax requires an additional after-tax to pre-tax translation factor.

THE IMPACT OF TAXES AND INTEREST RATES ON THE SERVICE PRICE

Estimates of the service price are now presented for U.S. corporate nonresidential depreciable investment under the current tax regime using formula (13.10b). (A more detailed description is contained in Appendix 13A.) Data for thirty-seven types of equipment and structures in seventy-three industries have been weighted using 1980 investment levels—the latest available official data—to construct average service prices for the economy. The real after-tax rate of return prevailing in 1983 is assumed equal across all industries. The nominal after-tax interest rate has been constructed to be consistent with the current 10 percent nominal long-term interest rate.

The estimates of the productive service streams (economic depreciation) were constructed using U.S. Department of Commerce methodology and U.S. Treasury studies which were used to estimate the 1962 Class Life System. The tax rates are estimates of the weighted average marginal tax rates in 1983 again using Treasury and Commerce information.

The service price of capital is influenced by technological factors, such as how quickly the capital wears out or becomes obsolete, by the tax structure, and by interest rates. Table 13—2 reports estimates of the service price of capital under various tax assumptions. The service price levels in percent represent the current *marginal value product of capital per unit of output* necessary to warrant undertaking the investment. The technology estimates were constructed by literally setting all tax parameters to zero. In a "no-tax" world, current technological factors are such that the marginal value product of capital per unit of output would be 18.99 percent for equipment, 6.04 percent for structures, and 14.61 percent overall.

The tax entries were similarly constructed by computing the service price with all other tax parameters set to zero. The results for property taxes, sales taxes, indirect business taxes (combined property and sales taxes), and direct business taxes (federal, state, and local income taxes) are shown separately. Taking all taxes into account, the service price for equipment increases from 18.99 percent to 25.16 percent, the service price for structures from 6.04 percent to 12.46 percent, and the service price for economywide capital from 14.61 percent to 20.86 percent.

The second set of entries in the table was constructed to illustrate the relative contribution of technological factors and taxes to the total service price. As can be seen, technological factors such as the productive life of a piece of capital comprise 70 percent of the economywide service price—the lion's share of the cost of employing capital. Because of nonlinearities, the entries are not additive.

Property taxes are the second largest contributor, accounting for 15.4 percent of the economywide service price. This result may seem surprising because property taxes are generally ignored in the analysis of capital costs. As explained earlier, their importance may be explained by the fact that the tax is applied to the "assessed value" of the capital asset whereas it is paid from the flow of revenue generated by the asset. Federal, state, and local income taxes follow next, contributing 9.4 percent of the service price, and sales taxes account for the smallest share, less than 4 percent.

The "EFFECTIVE TAX RATES" entry is the implied tax wedge on capital income. It is calculated by applying Eq. (13.10b). That is, the tax is equal to one minus the ratio of the "Technology only" line to the "All taxes" line. According to these estimates the *current federal, state, and local tax regimes increase the service price of equipment by 32 percent, of structures by 106 percent, and of economywide capital by 43 percent.*

As for the impact of interest rates, the estimate of the elasticity of the service price with respect to the real after-tax interest rate is shown below. It was constructed by increasing the nominal interest rate from 10.0 to 11.5 percent while holding the real rate of return to capital and expected inflation constant.

Elasticity of the Service Price of Capital
with Respect to the Real After-Tax Interest Rate

	Equipment	Structures	Total
Elasticity	0.0453	0.3701	0.1110

As can be seen, small changes in the real interest rate have *very little effect on the service price of capital and therefore on total investment.* A 10 percent increase in the real after-tax interest rate increases the service price of equipment by only 0.4 percent and the service price of structures by 3.7 percent. The effect of a 10 percent increase in the real after-tax interest rate on the service price of capital economywide is only 1.1 percent. Thus, the commonly held notion that movements in the real interest rate greatly affect the cost of investment, is not supported by either the geometry of investment decisions or the data used in their calculation.

In summary, the service price of capital depends on technological factors, interest rates, and taxes. This section has presented calculations which measure the impact of the various factors on the service price. The results indicate that, contrary to popular wisdom, changes in the real after-tax interest rate have little effect on the service price. The primary determinant of the service price is technology, a nonpolicy variable.

The other major determinant is the tax structure. Taxes, until recently, have been virtually overlooked in investment analysis. Some attention is now being turned to federal income taxes, in particular the treatment of depreciation. The results presented here, however, point to property taxes as the biggest contributor to the service price among the tax terms and second only to technology overall. These findings suggest that future analyses of capital costs need to examine the overall tax structure.

The preceding analysis of the real interest rate's influence on the real after-tax rate of return to capital is a partial one. A complete general equilibrium framework, not attempted here, would yield an even smaller impact of interest rates on the service price. First, the service price must equal the marginal value product of capital which cannot change very rapidly since it is based on the *entire stock of capital,* not merely new investment. This stock changes very slowly, replacing the depreciable portion every ten years on average. Short-term adjustments to fluctuations in the real interest rate would have to be translated into adjustments in the short-run demand conditions for the factors of production.

The appropriate demand schedule for the stock of capital (or investment) is its marginal value product schedule which relates the value of additions to output from increases in the use of capital. The elasticity of the marginal value product with respect to capital is equal to the factor's income share minus one. Using the standard estimate range of 3/10 to 1/3 for the income share of capital in the United States, estimates of the elasticity of the marginal value product schedule are in the −7/10 to −2/3 range, which means that stock changes would result in smaller (by 2/3) changes in magnitude in the service price.

Further, in the short run, an increase in the real rate of interest translates into a reduction in the real after-tax return to capital because the gross return, or service price, does not vary. The reduction in the net return to capital results in some substitution of labor for capital, which partially offsets the static change. This substitution alters the capital/labor mix at a lower level of capital. Thus, the final change in the service price as a result of a change in the interest rate would be smaller than those reported in the table. In the long run, the amount of investment would adjust to bring the return to real assets to its equilibrium value.

Finally, the impact of real after-tax interest rates on the service price of capital enters only through the income tax system. Higher interest rates reduce the present tax value of depreciation deductions since they are financial in nature. For the inframarginal, leveraged investments, the interest rate enters through the depreciation deductions and through the value of interest paid deductions. An increase in interest rates results in an increase in the service price of capital in the long-term but only through its impact on tax write-offs.

THE "REAL INTEREST RATE"

The level of real interest rates in the United States has been the subject of much attention lately. Figure 13–3 is a graph of the real after-tax interest rate and the real after-tax rate of return to capital (see Table 13–3). The latter measure was calculated rewriting Eq. (13.10b) such that the service price of each type of capital weighted by the appropriate stocks yields the observed gross return to capital in the United States. This process assures that the rate of return to each asset is equal and that the differential tax treatment of each is taken into account. As can be seen, the rate of return to real capital has been much more stable than the interest rate over the thirty-year period shown. The deviations largely occur during adjustment to new business conditions, such as changes in taxation.

In dramatic contrast, the real return to financial assets has experienced enormous swings during the period, the largest being the current positive swing. A similarly large negative swing occurred during the 1970s. What causes these swings? In the context of the framework developed before, they must be interpreted as swings in investors' expectations about future changes in inflation or taxes or in their evaluation of the riskiness of those forecasts. The return to real capital does not share this inflation sensitivity since its return is in real products which increase in nominal value with inflation. Thus, the difference in the nature of the risks facing financial versus real investments could explain the difference in the patterns of their real after-tax returns.

Others have argued that swings in the interest rate really represent changes in the "demand for credit." Although the precise definition of the actual market these analysts allude to is not clear, their general argument is that high government credit demands cause high real interest rates. Figure 13–3 clearly demonstrates that this analytical position is not borne out by empirical data. The third largest U.S. federal government deficit in history ($69 billion in 1975) occurred during the last *negative* swing. The government credit demand theory has not been supported by any econometric study of the empirical evidence to date. This is not to suggest that there is no relation between

Figure 13–3. Real After-Tax Returns to Capital and Financial Assets (1954–1983).

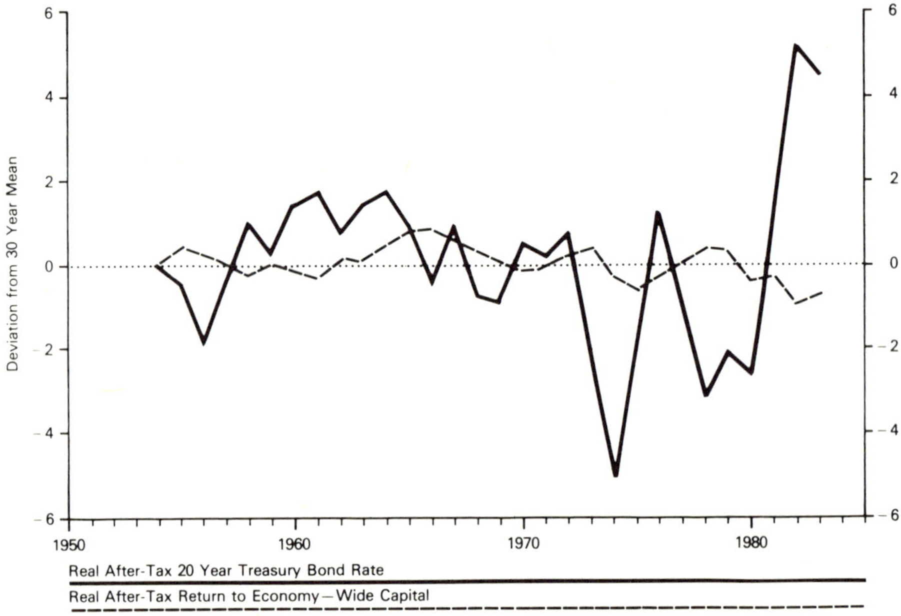

credit demands and interest rates but rather that large government credit demands cannot explain the large swings in real after-tax interest rates.

Figure 13–3 shows a continued decline in the real after-tax rate of return to capital through 1982 despite the "massive" 1981 tax cuts. This apparent anomaly may be explained by the fact that the property tax *rate* grew by 8 percent over 1981–1982. Given the major influence of this tax on the service price of capital, as previously shown, the real after-tax rate of return continued to fall until the personal tax cuts were phased in sufficiently to offset the increase in property taxes.

Table 13–3. Deviations from Thirty-Year Mean, Data for Figure 13–3.

Year	Real After-Tax Bond Rate (%)	Real After-Tax Return to Capital (%)
1954	−0.10	−0.06
1955	−0.49	0.34
1956	−1.93	0.19
1957	−0.53	−0.02
1958	0.96	−0.26
1959	0.27	−0.03
1960	1.42	−0.20
1961	1.71	−0.27
1962	0.74	0.06
1963	1.42	0.10
1964	1.72	0.43
1965	0.89	0.78
1966	−0.48	0.85
1967	0.89	0.52
1968	−0.75	0.31
1969	−0.92	0.07
1970	0.47	−0.21
1971	0.18	−0.09
1972	0.71	0.21
1973	−2.12	0.37
1974	−5.13	−0.28
1975	−1.72	−0.64
1976	1.14	−0.33
1977	−0.64	0.01
1978	−3.17	0.32
1979	−2.14	0.25
1980	−2.61	−0.45
1981	0.52	−0.28
1982	5.20	−0.94
1983	4.46	−0.76

Investment Implications

Whatever the causes of the current high real financial returns, many in the United States and the rest of the world are concerned that high real interest rates portend a downturn in economic activity. The high borrowing costs will choke investment incentives and end the current recovery. This argument is supposedly augmented by the fact that the federal government is also making large demands on credit markets which will further "crowd out" investment. The graphs in Figures 13—4 and 13—5 present data to test this proposition.

The graph in Figure 13—4 plots an investment rate defined as the deviations from the thirty-year mean of the ratio of business fixed investment to total gross national product against the real after-tax return to financial assets. (See Table 13—4.) Under the hypothesis stated earlier, the after-tax borrowing cost to an investor in real assets should closely mirror the similar return to a lender. Investment, therefore, should be highly correlated with the after-tax return to financial assets. This does not, however, seem to be the case. During the 1970s and 1980s there appears to be little relation between the two series. In fact, the United States is currently enjoying a capital boom in spite of the historically high real rates. This graph tends to confirm the result reported earlier in the chapter that the impact of borrowing costs is a minor one in the translation of the purchase price of capital goods into the required service price.

The graph in Figure 13—5, which plots the rate of return on real assets against the investment rate clearly indicates that the real rate of return to capital as derived from the service price is a much better predictor of investment behavior than real financial interest rates. Large swings in the real after-tax interest rate do not coincide with immediate adjustments in investment. Rather, the swings occur at turning points in the level of investment activity. The direction of influence seems more likely one of the profitability of alternative real investments affecting the required financial return rather than financing costs affecting investment. This corresponds closely with the classical notion that the interest rate is made up of the return to real capital plus a premium for risk plus an inflation premium.

The proposition that income could be redistributed at little cost to aggregate economic activity was based on early empirical research that indicated little or no responsiveness of investment to the rate of return. Since the rate of return on investment was mistakenly assumed to be measured by the real interest rate, the graph in Figure 13—6 indicates how one could reach that conclusion. As can be seen from this graph and Table 13—5, there is little correlation between the real interest rate and investment, and a statistical regression analysis would show extremely small or insignificant coefficients between the two. The real interest rate series is the same twenty-year Treasury bond rate less

Figure 13–4. Real Financial After-Tax Return and Rate of Investment (1954–1983).

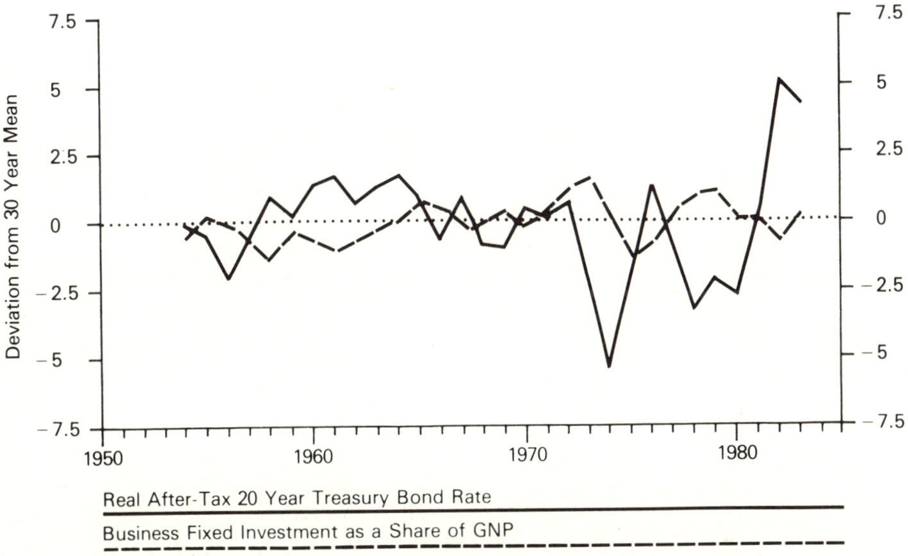

current inflation *but before taxes.* The investment rate is the ratio of business fixed investment to gross national product. Both series are again presented as deviations from their thirty-year means, but the series have been scaled to their largest positive value to simulate the mathematics of a linear regression. The omission of taxes greatly misstates the return to investment, and the best characterization of the relation of the two series would be a random walk.

In the graph in Figure 13–7, the "price" term is the real *after-tax* return to capital as defined previously which corrects the mismeasurement problem in the graph in Figure 13–6. The investment rate has been calculated as the ratio of investment to private output available to be allocated by the private sector. The private output measure is GNP less the goods and services preempted by the government and net exports. The same scaling method used in Figure 13–6 is applied here. The "fit" is quite striking and certainly does not support the

Figure 13–5. Real After-Tax Return to Capital and Rate of Investment (1954–1983).

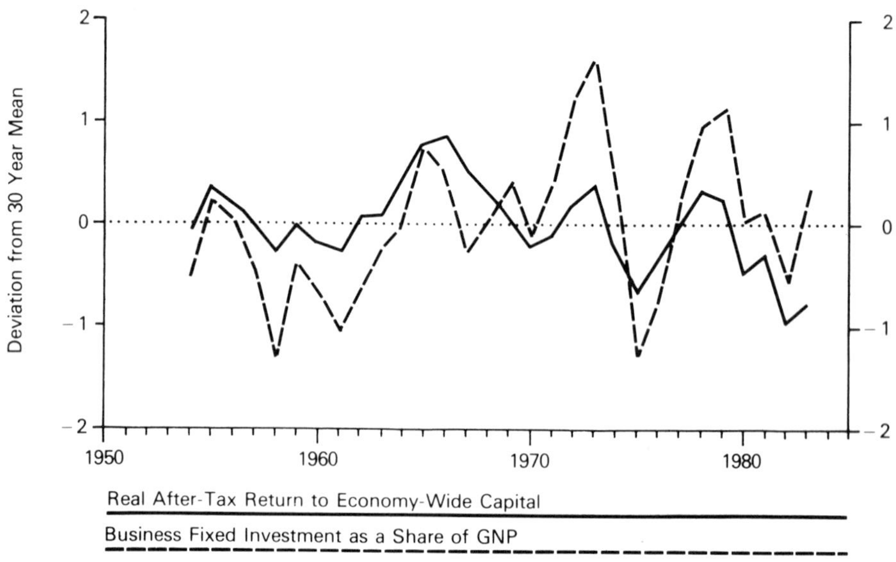

conclusion that the two series are unrelated.

These two charts support the previous conclusion that the real interest rate has a relatively small influence on both the service price of capital and the rate of return to capital. Figure 13–6 should be no surprise, and Figure 13–7 is simply as classical theory would predict. The lesson to be learned is that the effect of taxes must be taken into account when relative prices are called for by the theory. A similar mistake is made in current econometric practice when an assumption is made that the interest rate is the appropriate discount rate for calculating the service price. This brings us back to the graph in Figure 13–5, which is clearly an inferior relation. Finally, the level of government preemption must be subtracted from the "income term" in the relation to gauge correctly the budget constraint in the allocation of products between consumption and investment.

Table 13–4. Deviations from Thirty-Year Mean, Data for Figure 13–4.

Year	Real After-Tax Bond Rate (%)	Business Fixed Investment as a Share of GNP (%)
1954	−0.10	−0.53
1955	−0.49	0.23
1956	−1.93	0.04
1957	−0.53	−0.42
1958	0.96	−1.26
1959	0.27	−0.39
1960	1.42	−0.66
1961	1.71	−1.04
1962	0.74	−0.67
1963	1.42	−0.26
1964	1.72	−0.02
1965	0.89	0.70
1966	−0.48	0.47
1967	0.89	−0.27
1968	−0.75	0.04
1969	−0.92	0.37
1970	0.47	−0.12
1971	0.18	0.40
1972	0.71	1.20
1973	−2.12	1.59
1974	−5.13	0.37
1975	−1.72	−1.27
1976	1.14	−0.76
1977	−0.64	0.29
1978	−3.17	0.96
1979	−2.14	1.10
1980	−2.61	0.05
1981	0.52	0.14
1982	5.20	−0.54
1983	4.46	0.26

Figure 13–6. Real Interest Rate and the Rate of Investment (1954–1983).

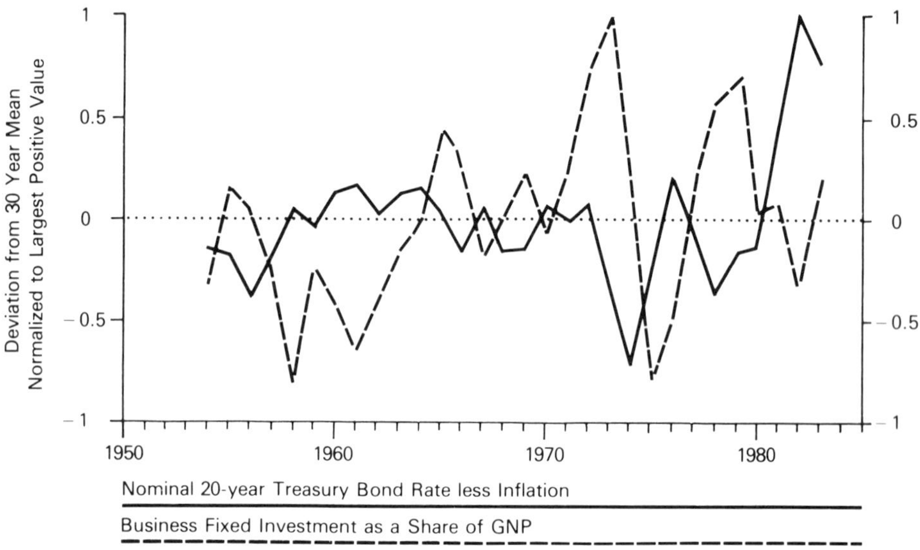

CONCLUSIONS

Where does this leave us in the discussion of the service price of capital and what does this type of analysis tell us about likely future U.S. policy adjustments? The clear message from the graphs is that the high "real interest rate" is a misleading indicator of current economic incentives to make real investments. Rather, it is the rate of return on real assets, as determined by the service price of capital, which determines the level of business investment. A lower required service price leads to greater investment and higher economic growth. Indeed recent reductions in the service price of capital were made possible in part by the Economic Recovery Tax Act of 1981 (ERTA) which lowered marginal tax rates and instituted an accelerated cost recovery system of depreciation. The full impact of ERTA tax cuts has only become fully effective during the last year.

Table 13–5. Deviations from Thirty-Year Mean Normalized to Largest Positive Value, Data for Figure 13–6.

Year	Bond Rate Less Inflation (%)	Business Fixed Investment as a Share of GNP (%)
1954	−0.3306	−0.1411
1955	0.1437	−0.1879
1956	0.0247	−0.3866
1957	−0.2637	−0.1706
1958	−0.7912	0.0420
1959	−0.2455	−0.0340
1960	−0.4132	0.1273
1961	−0.6501	0.1638
1962	−0.4211	0.0244
1963	−0.1628	0.1233
1964	−0.0103	0.1535
1965	0.4383	0.0273
1966	0.2925	−0.1544
1967	−0.1707	0.0554
1968	0.0256	−0.1525
1969	0.2295	−0.1422
1970	−0.0766	0.0655
1971	0.2479	−0.0061
1972	0.7550	0.0687
1973	1.0000	−0.2991
1974	0.2350	−0.6954
1975	−0.7944	−0.2028
1976	−0.4794	0.2073
1977	0.1796	−0.0525
1978	0.6037	−0.3471
1979	0.6919	−0.1613
1980	0.0334	−0.1288
1981	0.0887	0.4190
1982	−0.3415	1.0000
1983	0.1615	0.7847

The analysis suggests that a tax increase for the purpose of reducing the deficit in order to lower interest rates and promote investment is completely misdirected. The tax increase would increase the cost of labor and capital, thereby reducing the growth rates of investment and

Figure 13–7. Real After-Tax Return to Capital and Investment's Share (1954–1983).

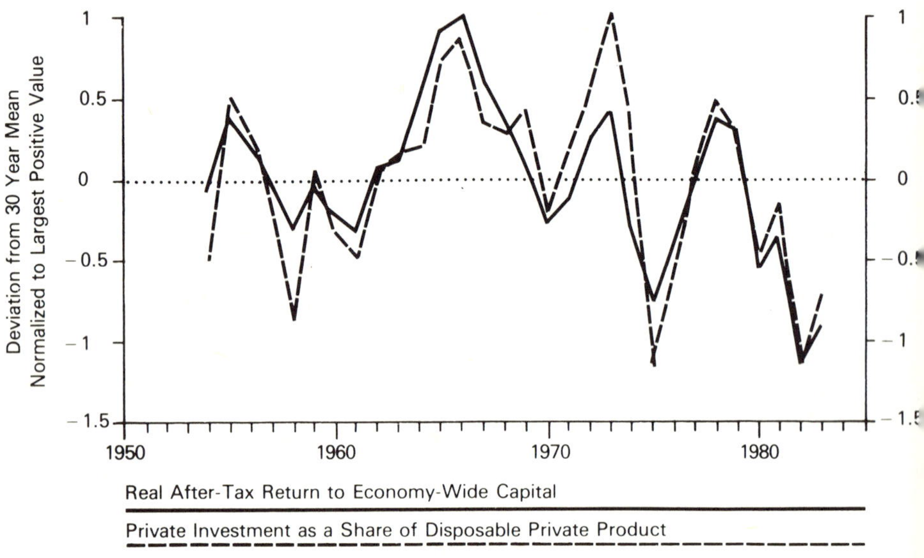

output. Any decline in interest rates would reflect the fall in investment and GNP growth. Slower growth could prompt greater inflation, or fears of inflation, and result in higher rates of interest.

During the recent election campaign President Reagan repeated his commitment to long-term economic growth. This represents a fundamental shift in the approach of the federal goverment to the economy. Over the last four decades the United States had been primarily influenced by a welfare state philosophy. Support for this philosophy was made possible through use of an income-expenditure macroeconomic framework, which held that the government could costlessly redistribute income among individuals via the tax and transfer system. Under this approach transfer payments to individuals grew from 14.7 percent of government expenditures in 1954 to 31.6 percent in 1983, or an increase of 216 percent. The taxes needed to support this increase grew as well.

The 1980s seem to be ushering in a new approach to government and the economy. The ideal of a welfare state is being challenged by the concept of an opportunity society. This approach recognizes that the transfer of income is not costless, particularly when done through the tax and welfare system. Rather than redistribute existing wealth, the emphasis is on creating new wealth through growth. To some extent this means starting fresh from where we are today while making sure that in the future everyone has the same opportunity to make the best use of his or her individual talents.

Continuing growth is predicated on increasing productivity, which, in turn requires increasing the capital-to-labor ratio. If the United States is not successful in its attempts to achieve long-term growth through increased investment and higher productivity, the current adjustments underway in the older so-called smokestack industries to restore their competitiveness in world markets will be undercut. Protectionist pressures in the United States would increase, and the Western Alliance would unravel as the United States closed its markets to its allies. Such prospects are so unattractive that we believe a supply-side policy will prevail.

APPENDIX 13A: DESCRIPTION OF SERVICE PRICE CALCULATION

In order to evaluate alternative tax regimes, it is necessary to measure their impact on the cost of capital services. An increase in these costs would require an increase in the gross returns required by acceptable investment opportunities and, hence, a decrease in the amount of real investment undertaken. This, in turn, would mean a lower capital stock, a less productive labor force, and a lower GNP. We have measured the alternative costs of capital implied by various depreciation proposals via a "service price" calculation for each of thirty-seven different asset categories covering seventy-three different industry classifications. This appendix describes those calculations.

The service prices calculated for each asset category represent the current marginal products required per dollar of corporate investment in that asset by each industry. They are the before-tax rates of return required to be produced by the asset in order that the anticipated taxes, depreciation, and a "normal" rate of return are covered. The normal real rate of return is assumed equal to 2.7 percent, a level we estimate to have prevailed during 1983. An asset category's rate of *economic* depreciation is assumed generally to vary across industries. Allowable *tax* lives also generally differ across industries, and allowable depreciation methods vary among the several alternative tax regimes in place in the United States during the period 1954 to 1983. These regimes include

1. Bulletin F Guideline Lives
2. Class Lives, using Asset Depreciation Range (ADR) write-off methods.
3. ADR, using that life within the given range that minimizes the service price (accounting for different investment tax credit rates according to the chosen depreciable life).
4. Accelerated Cost Recovery System (ACRS) as originally passed in 1981 under ERTA
5. ACRS as currently implemented

The algebraic expression used for the calculation of the service price is derived from the first-order condition for a wealth-maximization problem. It assumes a constant investment deflator, measured relative to an overall price deflator, is expected to prevail over the relevant future. The maximization calculus is performed from the standpoint of the ultimate investor, the individual stockholder. The existence of a corporate legal structure is deemed important only insofar as it creates an additional tax liability for the investor. The alternative—neglecting taxes on dividends—would be unsatisfactory: a corporation that (either explicitly or implicitly) neglects the additional taxes on dividends would fail to provide its stockholders with a market level, after-tax rate of return.

The service price expression for corporate capital is given as

$$\text{Service price} = tp/(1 - ts) + (1 - k - D)/t(1 - ts) \times((1 - tcf)\times(1 - td)\times(1 - tcs))\times E], \tag{13.11}$$

where

tp = the rate of property taxation, assumed equal to 3.22 percent for current law examples.

ts = the rate of tax on output (for example, a value-added tax or sales tax on all final product), assumed equal to 5.26 percent.

tcf = the federal corporate tax rate, assumed equal to 46 percent for current law examples.

tcs = the state corporate tax rate, assumed equal to 9.67 percent.

td = the marginal dividend tax rate, adjusted to reflect the average timing difference between profit accruals and dividend receipts, assumed equal to 16.37 percent initially. This rate is assumed to be subject to bracket creep for future periods.

k = the effective rate of the investment tax credit (the statutory credit, adjusted for the net income limitation), assumed equal to 9.23 percent for equipment and nonbuilding structures. This rate is adjusted downward for short-lived assets under Class Lives, ADR, and ACRS.

D = the present value of the future stream of tax depreciation allowances, adjusted to an after-tax basis; that is, the stream is multiplied by an appropriate tax factor. A nominal interest rate of 6.7 percent is used as the discount factor, reflecting a 4 percent rate of inflation assumed throughout.

E = the present value of the "efficiency stream," the present value of the future real returns, measured as a percentage of the asset's initial marginal contribution to output. A real interest rate of 2.7 percent is used as the discount factor. A further description follows.

The assumed average *economic* life for each asset/industry category is the applicable Class Life under the old ADR system. These lives were first introduced in 1962 under the Guidelines depreciation system. In some cases, the Bureau of Economic Analysis (BEA) asset categories that were used do not correspond exactly with the IRS Class Life categories. In those cases an average or representative life was chosen.

Variation in expected asset lives is simulated by the use of a truncated normal distribution centered on the assumed average economic life. This distribution is used to derive an asset "discard" function. The discard function asumes that some (small) proportion of an original investment in assets of a certain type is discarded beginning at 50 percent of the assumed average economic life. It also assumes that some (equally small) proportion of the original investment is maintained up to 150 percent of the assumed average life. The other capital assets constituting the original investment bundle are discarded at ages in between 50 percent and 150 percent, with the greatest number being discarded at the average economic life. In addition, a concave efficiency function is assumed for all nondiscarded assets to reflect factors such as technological change. This function assumes that the loss of productive efficiency is smallest in the early years, and greatest in the final years for each particular asset. (This is the reverse of a geometrically declining efficiency schedule in which the greatest absolute efficiency losses are incurred immediately.) Combining the discard function with the concave efficiency function yields an overall efficiency function for a given investment bundle. The general shape of this function indicates an initial slow rate of efficiency loss for the investment, a faster rate as the original investment ages and assets are discarded, but again a slower rate as we reach the upper tail of the discard function. The overall function becomes zero at 150 percent of the average economic life. This methodology is identical to that used by the Office of Business Analysis, U.S. Department of Commerce in generating their capital stock database.

The variable E represents the present value of the efficiency function just described. Alternatively, it can be viewed as a measure of the average life of a given asset category expressed in units of current output. In the absence of taxation, the inverse of E by itself, would represent the "cost of capital." For example, with an infinitely lived asset (no efficiency loss and no discards), E would equal the present value of an infinite series of ones, or simply 1 over the discount rate. The inverse of E would therefore equal the real rate of interest. Similarly, under an assumption of geometrically declining efficiency, the inverse of E would equal the sum of the interest rate and the (constant) rate of depreciation. With the efficiency schedule described, however, the present value formula cannot be so easily condensed, and the more general form of the inverse of E must be used to measure the joint requirement for interest and economic depreciation.

The *tax* depreciation write-offs used in deriving the D term are calculated according to the relevant taxation scheme. In all cases, a half-year convention is used. The appropriate tables found in the tax regulations are used for calculating personal property allowances under ACRS. A choice is allowed whereby either the original depreciable basis is adjusted downward by 50 percent of the Investment Tax Credit or a 2 percent reduction in the allowable credit is taken. The allowance schedule for eighteen-year real property was constructed using the 175 percent declining balance method. Under the ADR and Class Life Proposals, either a declining balance method (with a switch to straight line at the appropriate time) or sum of the years' digits method is chosen. For Section 1245 property, either 200 percent declining balance or the sum of the years' digits method is chosen. For Section 1245 property, either 200 percent declining balance or the sum of the years' digits method is used, depending on which yields the greatest present value of depreciation allowances. For Section 1250 property (which we equate with BEA's building categories), the method is limited to the 150 percent declining balance method.

The straight-line method of depreciation is used in calculating taxable dividends under all regimes. For ACRS, the lives specified in Code Section 312(k) are used. A five year write-off is used under the expensing alternative. Inflation indexing is not taken into account in this calculation.

The tax rate on dividends is assumed to increase slowly over time due to bracket creep. Both a real growth adjustment and an inflation adjustment in the marginal rate is incorporated. Possible anticipated future tax "cuts" intended to correct for bracket creep are not taken into account. The real growth factor used in the calculations is 2 percent; inflation is assumed to be 4 percent. Bracket creep elasticities of 0.3 are used, so that a 1.8 percent $(0.3(0.02 + 0.04))$ annual increase in dividend tax rates is assumed. This translates into roughly a 30-basis-point increase in the dividend tax rate per year. A dividend tax rate ceiling of 50 percent is imposed.

The nominal depreciation allowances are multiplied by the appropriate tax rates in order to express their impact in after-tax terms. The D term is defined by the following expression, which accounts for the deductibility of state corporate taxes on the federal return:

$$D = \text{Present value}[(tcf + tcs - tcs(tcf + td)) \times \text{Corporate allowance} + (td) \times \text{Straight line allowance}] \qquad (13.12)$$

This D term is akin to the investment tax credit as regards its impact on the cost of capital. The depreciation allowances reduce the cost of the initial investment in present value terms by a percentage equal to D. Notice that there is no necessary connection between D, which is based on the allowable *tax life*, and E, which is based on the assumed distribution of *economic lives*. Equating the tax life with an average economic life does not necessarily yield a more or less burdensome tax system. Also, varying the assumptions regarding the economic lives and the pattern of efficiency decay does not change the relative rankings of alternative depreciation rules on the service prices.

NOTES

1. See Keleher and Orzechowski (1980) for a history of supply-side thought.
2. A change in the price of debt will alter the equilibrium debt/equity position of the firm, changing its degree of risk. This will be reflected in the real rate of discount which is being held constant for this analysis. This is not an unreasonable assumption in light of the economywide accounting system also being used.
3. An investment tax credit granted at the time the investment is put into service has an immediate direct tax effect; for example, a 10 percent credit will offset tax liability up to 10 percent of the asset's purchase price. (Excess credits are normally carried forward until exhausted.)

REFERENCES

Keleher, R.E., and W.P. Orzechowski. 1980. "Supply-Side Effects of Fiscal Policy: Some Historical Perspectives." Federal Reserve Bank of Atlanta Working Paper Series. August.

U.S. Treasury Department, Office of the Assistant Secretary for Economic Policy. 1984. *The Effects of Deficits on Prices of Financial Assets: Theory and Evidence.*

14 THE TREASURY TAX REFORM PLAN

Charls E. Walker

All business owners and managers should strongly support the U.S. Treasury Department's tax reform goals of fairness, simplicity and lower rates for individuals and business. Many businesspeople are, however, deeply concerned about the means advanced to finance these objectives.

That proposal would fund reductions in tax rates primarily by raising taxes on corporations by $165 billion over the next five years. By 1990, when the Treasury plan would be fully effective, corporate tax payments would be 37 percent higher than today.

As President Reagan has said many times, corporations do not pay taxes, people do. Any tax on a corporation is inevitably passed on to people—the corporation's customers, owners or workers. We do not know to what degree the corporate tax is passed on to each of these groups. It is therefore a hidden tax in the worst sense of the word.

But that's not all. To the extent the proposed corporate tax increase is passed on to consumers through higher prices, it is doubtless regressive in impact. This is because people with low incomes spend a larger portion of those incomes on the products of U.S. business corporations than do people with high incomes.

Moreover, it is simply common sense that high taxes on business hurt jobs, growth, and productivity. Money that business corporations pay to Uncle Sam in taxes cannot go into paychecks, nor can it be used to expand or modernize plant and equipment. Since much of any increase in productivity comes from installation of state-of-the-art equipment and machinery, higher business taxes impede the growth in efficiency that is so important to growth and to U.S. competitiveness in world markets. With the U.S. trade deficit at $123 billion per year, the bolstering of such competitiveness should be high on the agenda of national policy.

TREASURY'S REJECTION OF ACRS AND ITC

The antigrowth impact of the increase in corporate taxes in the Treasury plan is magnified significantly by the nature of the increase, for Treasury would repeal those business investment provisions which are most conducive to—and in fact were designed especially to stimulate—productive capital formation. These provisions are the Investment Tax Credit (ITC) and the Accelerated Cost Recovery System (ACRS) which, ironically enough, formed the business centerpiece of the Reagan administration's 1981 supply-side tax cut. (It is important to note that Ronald Reagan's 1981 tax proposals contrast sharply with much of the current Treasury plan. Along with the rebuilding of the nation's defenses, the 1981 tax policy victories stand out as a signal and much-to-be applauded achievement of Ronald Reagan's first presidential term.)

The ACRS provisions are not Johnny-come-lately; they have been around for a long time and own highly respectable pedigrees. Accelerated depreciation was the central provision of a supply-side tax cut pushed by President Dwight D. Eisenhower in 1954. In 1961, President John F. Kennedy proposed enactment of the investment tax credit as the centerpiece of the Kennedy—Johnson supply-side tax cuts.

How explain the fact that a Treasury which works for a dedicated supply-side president and which itself strongly pushed for ACRS/ITC in 1981 now calls, not just for its modification, but for outright repeal? Treasury eschews analysis of the supply-side triumphs that ACRS/ITC has achieved to date and may well achieve in the future; instead, it advances two more or less theoretical arguments: First, the differential tax rates that ACRS helps engender are unfair. (Treasury presumably believes that corporate tax "fairness" requires that all firms and industries pay identical effective tax rates.) Second, tax breaks for capital-intensive companies lead to "an uneven playing field." (Treasury maintains that this causes distortions in resource allocation that would not occur if special tax breaks were forgone and resource allocation were left to the market.)

A third argument for at least modifying ACRS and ITC is advanced by some members of Congress, namely, that the 1981 investment tax breaks did not receive adequate congressional consideration in the first place, and turned out to be far too generous (even though they were cut back by 38 percent in 1982).

Note, however, that the heavy business tax increases in the Treasury plan do not, as some administration spokesmen contend, represent only increases in taxes on companies that pay little or no taxes now. Treasury estimates that repeal of ACRS and ITC would raise a whopping $300 billion in new revenues in the next five years. $300 billion in new revenues *cannot* be attributed solely to tax leveling.

CORPORATIONS: THE SURROGATE TAX COLLECTORS

Take first the Treasury argument that the existence of differential effective tax rates among firms and industries is "unfair." Unfair to whom? Fairness in taxation cannot be judged in terms of impact on institutions that people have created to help meet their economic needs. Fairness can only be judged in terms of the impact of taxes on people themselves. Levying a tax on corporations, which are only legal arrangements for doing business (and very good ones at that), is only an intermediate step; in effect, a corporation is simply a surrogate tax collector for IRS. To be meaningful, judgments as to the equity of any tax must be based on analysis of its impact on the true taxpayers—people—not on institutions. The trouble is that there is no general rule that tells us how these hidden corporate taxes are ultimately distributed among people.

Consider the impact of the Treasury plan on two important industries on which taxes would be raised, steel and electric utilities. A worldwide surplus of steel guarantees that, absent an increase in tariffs or quotas, higher taxes on steel companies in the United States cannot be passed on to buyers; if U.S. producers raise prices, foreign producers will capture the market. The steel tax increase must therefore be borne by workers and owners—the former through lower pay, shorter hours or even loss of jobs; the latter through lower returns on their investment. Is that fair?

As to electric utilities, regulatory authorities permit all of their taxes to be passed through to consumers in the form of higher rates. Since almost everyone buys electricity, the Treasury's proposed tax increase on electric utilities is a blatantly regressive tax on people. Is that fair?

AN UNEVEN PLAYING FIELD

Treasury's second major argument for repealing ACRS is that investment tax breaks create "an uneven playing field"; allocation of resources is distorted and economic efficiency is impaired. This argument might have some merit if we thought only of the present and ignored the future, but this we can do only at our own peril and that of future generations. Resource allocation has an important time dimension: How much of our resources are consumed now? How much in the future? For consumption to be shifted to the future, saving and investment are necessary today. And the fact is that any income tax is inherently biased against the saving and investment that is necessary to shift consumption from present to future.

Income taxes on individuals hit saving at least twice—once when the income is earned, again when the income on the invested savings is received (recognition of this led Congress to create Individual Retirement Accounts, thereby making it possible for most individuals to eliminate all or a substantial part of the taxes on their savings). The problem is even worse with respect to business corporations, *where saving can be taxed at least three times*—first when the corporate income is earned, again when the income on corporate saving and investment (in the form of new plant and equipment) comes in, and a third time when taxable stockholders receive their dividends.

All taxation of corporate saving could be eliminated if the corporate income tax were to be repealed, a view that has gained strong support in the academic community. Note the views of two leading academic economists. Conservative Arnold Harberger (University of Chicago) says that the corporation income tax "has no sound economic underpinning." Liberal Lester Thurow (Massachusetts Institute of Technology) says, "the corporate income tax should be abolished regardless of whether you are a conservative or a liberal." Politically, however, repeal is not in the cards. The next-best-thing would be to reduce one big layer of taxation of corporate saving by providing for "expensing" (immediate write-off) of business capital assets. Indeed, Treasury tax experts now appear to agree with this view, but state that, absent a shift to consumption taxes in the United States (something the Treasury rejected in its study, and regrettably so), technical problems of tax policy create problems for this approach.

We disagree. In fact, the 1981 enactment of ACRS and liberalization represented an effort to approximate expensing for equipment, and won out in the House debate on the bill over an outright expensing proposal pushed by Democrats on the Ways and Means Committee. ACRS and ITC are especially attractive as a means of reducing triple taxation of corporate saving because they automatically direct a big part of untaxed saving into the plant and equipment that support jobs, growth and international competitiveness. Indeed, ACRS and ITC provide the biggest supply-side "bang for the buck"; a company gets the tax breaks only if it makes qualified investment in plant or equipment. There is a minimum of leakage.

Rather than representing subsidies to investors, ACRS and ITC reduce excessive taxation of corporate saving and investment, and provide the most productive route to a truly level playing field. The Treasury plan, by wiping out ACRS and ITC, would make the playing field even more uneven—that is, biased against saving and productive investment—than it is today.

ACRS/ITC: A CAREFULLY CONSIDERED PROPOSAL

Was the 1981 business tax cut an overly generous measure that received inadequate congressional consideration?

Consider first, the origin of ACRS/ITC and its congressional support. ACRS/ITC was the legislative adaptation of the well-known "10—5—3" proposal for business depreciation. It was developed in early 1979 by a group of business tax experts and lobbyists under the policy guidance of congressional Representatives James Jones and Barber Conable, two highly respected members of the tax-writing Ways and Means Committee. When they introduced 10—5—3 as the Capital Cost Recovery Act in the summer of 1979, they were joined by highly respected members of the Senate Finance Committee, including current Chairman Bob Packwood and second-ranking Democrat Lloyd Bentsen of Texas.

The 10–5–3 was most attractive to a Congress increasingly concerned about lagging productivity growth in the United States as contrasted with our major competitors abroad. Within a short time, more than three hundred members of the House of Representatives and more than seventy senators had signed on as cosponsors of the proposal.

ACRS/ITC was not foisted upon anyone. It was not, as some critics charge, a product of secrecy and stealth in the late hours of Congress. It was the product of highly respected members of Congress' tax-writing committees. And it was on the congressional table for a full two years—with overwhelming support—before its final passage in August 1981.

TOO MUCH OF A GOOD THING?

Regardless of origin and support, is ACRS/ITC too generous, even after substantial cutback in 1982? It has become fashionable to say so, but again the record suggests otherwise.

The question of whether ACRS/ITC is "too rich" can only be answered in terms of what Congress and the Reagan administration were trying to achieve in enacting it in the first place. Complex and esoteric equations relating capital cost recovery systems to "present value" and similar calculations are not arguments in the real world where politicians and people live and work; they are more suited to college exam questions in theoretical economics.

The Reagan administration recommended and Congress enacted ACRS/ITC for one simple reason: deep concern over the low rate of capital formation and productivity growth in the United States and a conviction that a complex and too restrictive capital cost recovery system was the major culprit. It therefore follows that ACRS/ITC can be said to be too liberal *only* if we are now getting a higher rate of business fixed investment than we want.

ACRS/ITC is doing what its proponents said it would do and what Congress and the administration wanted it to do. Despite interest rates that are still high by historical standards, business fixed investment led the recent economic recovery and, in fact, has been the strongest in any recovery since the late 1940s (when long-term Treasury interest rates were only 2.5 percent). Analysts in the Congressional Budget Office, at Wharton Econometrics, and elsewhere are concluding that the current strength of business investment can be traced to a combination of the drop in the inflation rate and the enactment of ACRS/ITC, which increased both the after-tax rate of return on new investment and the cash flow that businesses need to finance that investment.

The high rate of business investment has not only led and helped sustain the business recovery; it has also contributed to anti-inflationary increases in plant capacity. This is estimated at 2 to 2.5 percent in both 1983 and 1984, and is expected to make a big jump of 4 percent in 1985.

In addition, the new business investment is concentrated in modern, efficient, state-of-the-art equipment. This will help greatly to increase U.S. productivity—and hence international competitiveness—in the years ahead.

In enacting ACRS/ITC, Congress and the Reagan administration were intentionally trying to change resource allocation, to restrain consumption and enhance capital formation. Congress succeeded. ACRS/ITC as it applies to industrial plant and equipment is working beautifully.

"If it ain't broke, don't fix it."

TRADE AND THE COST OF CAPITAL

The huge increase in hidden and regressive taxes and the repeal of ACRS/ITC are not the only disturbing aspects of the Treasury tax plan. By raising maximum taxes on individual capital gains from 20 to 35 percent and in corporate capital gains from 28 to 33 percent, Treasury would in effect reverse one of the most successful tax cuts in history. The reductions in tax rates on capital gains have fostered a surge in venture capital, bolstered equity values, and enhanced rather than reduced federal revenues.

When the increase in tax rates on capital gains is added to repeal of ACRS and ITC, the international dimensions of the Treasury proposal begin to come into focus. Together, and even after allowing for the cut in corporate tax rates to 33 percent, indexing of capital gains, and some deductibility of dividends, these changes would increase the cost of capital to American business. Absent a significant drop in interest rates (which depends more on deficit control and monetary policy than tax policy), U.S. firms will find themselves increasingly behind the 8-ball in world markets.

For example, the cost of capital in this country is already considerably higher than in Japan, one of our principal competitors. Costly capital impedes the investment that is the key to productivity growth and more efficiency in output. Given the U.S. trade deficit, which is both huge and frightening, sensible public policy calls for a reduction, not an increase, in the cost of capital in the United States.

TREASURY PLAN'S TAX INCREASES: HIDDEN, REGRESSIVE, AND ANTIGROWTH

Given the case against the corporate income tax, can any policy that contemplates a 37 percent increase in those taxes by 1990 be considered fundamental tax reform? Is shifting of the tax burden from individuals to corporations (or, as Ronald Reagan himself might point out, shifting of burdens on some individuals in a hidden manner to other individuals) the essence of fundamental tax reform? Many academic experts would say it is not; to them, the corporate income tax makes no sense.

Finally, consider this. If enacted as proposed, the Treasury tax plan would close out a decade of major (and often contradictory) Federal income tax changes. In 1990, the final scorecard would show *a net cumulative increase* of income taxes on corporations of $115 billion and a net cumulative decrease of income taxes for individuals of $600 billion.

Given those figures, historians would *not* view the 1980s as a period of fundamental tax reform in the United States. To the contrary, they would wonder why our generation of Americans suddenly reversed a sensible progrowth tax policy, that was working well, and shifted instead toward taxes that are hidden, regressive and antigrowth.

INDEX

ABOUT THE CONTRIBUTORS

Alan J. Auerbach is professor of economics at the University of Pennsylvania. He previously taught at Harvard University and visited Yale as the Irving Fisher Research Professor. He is a research associate at the National Bureau of Economic Research where he directs the bureau's Project on Mergers and Acquisitions. He has published numerous articles in scholarly economics and law journals and serves as associate editor of the *Journal of Public Economics.* On matters of tax reform, he has appeared frequently as a witness before congressional committees. He served as a consultant to the U.S. Treasury during the formulation of its own tax reform proposal, introduced in November 1984.

Carlos E. Bonilla joined the Institute for Research on the Economics of Taxation (IRET) in the summer of 1984. At IRET he has authored numerous studies dealing with such diverse matters as the Section 936 tax credit for possessions corporations and tax provisions for capital recovery. Prior to joining IRET he was an economist in the Office of Income Security Policy in the Department of Health and Human Services. Mr. Bonilla holds a master's degree in economics from Georgetown University.

Michael J. Boskin is professor of economics, chairman of the Center for Economic Policy Research at Stanford University, and research associate, National Bureau of Economic Research. He is the author of approximately fifty articles and editor of six volumes of essays on taxation, fiscal policy, capital formation, labor markets, social security, and related subjects. He has been a guest columnist for (or featured in) the *Wall Street Journal*, the *New York Times*, the *Los Angeles Times, Newsweek, Time, Fortune, Forbes*, and *Business Week.* He received his B.A., M.A., and Ph.D. degrees from the University of California, Berkeley. He has been a consultant and frequent witness to the committees of Congress dealing with economic policy and to the Treasury Department. He was a member of several of President Reagan's economic policy task forces during the president's 1980 campaign.

Stephen H. Brooks is an independent economic consultant specializing in the design and implementation of economic information systems. These activities include the development of economic modeling and analytic tools to be used for policy analysis by government agencies and for budgeting, planning, and strategic analysis by businesses. He was senior staff economist on the President's Council of Economic Advisers, senior economist at Data Resources, Inc., and has taught at the University of Michigan. He received his A.B. in economics from Harvard and his Ph.D. from the University of Michigan.

Benjamin M. Friedman is professor of economics at Harvard University, where he teaches macroeconomics and monetary economics and conducts a seminar on monetary and fiscal policy. He received A.B., A.M., and Ph.D. degrees in economics from Harvard University. In addition, he received an M.Sc. in economics and politics from King's College, Cambridge, England, where he studied as a Marshall Scholar. He is the author of numerous books and articles. He currently serves as director of financial markets and monetary economic research at the National Bureau of Economic Research, as a director of the Private Export Funding Corporation, and as an associate editor of the *Journal of Monetary Economics.*

George N. Hatsopoulos is the founder, chairman of the board, and president of Thermo Electron Corporation, a company whose principal business is the development and manufacture of process equipment and instruments for energy-intensive industries. Dr. Hatsopoulos received his education at the National Technical University of

Athens and the Massachusetts Institute of Technology where he received his B.S., M.S., M.E., and Sc.D. degrees. Dr. Hatsopoulos served on the faculty of M.I.T. from 1956 to 1962 and has continued his association with the institute, currently serving as senior lecturer. He is a member of the National Academy of Engineering and received the Pi Tau Sigma Gold Medal Award for Outstanding Achievement in the Field of Engineering for the years 1950 to 1960.

Edward G. Jefferson is chairman and chief executive officer of the DuPont Company. He is a cochairman of The Business Roundtable; a member of the American Section of the Society of Chemical Industry, of which he was formerly chairman; and a vice chairman of The Conference Board. He also serves on the Board of Trustees of the United States Council for International Business and is a member of the Executive Committee of the Private Sector Council.

Burton J. McMurtry is one of the founding general partners of Technology Venture Investors, a series of venture capital partnerships begun in 1980. He entered the venture capital business in 1969 and was president of Palo Alto Investment Company from 1970 to 1973. In 1973 he was one of the founding general partners of Institutional Venture Associates. Before entering the venture capital business he spent twelve years in engineering and management positions during which time he authored more than twenty technical publications in professional journals and books and was elected a fellow of the Institute of Electrical and Electronic Engineers. He holds B.A. and B.S.E.E. degrees from Rice University and M.S. and Ph.D. degrees in electrical engineering from Stanford University. He is an officer and director of the National Venture Capital Association and has served as president and as a director of the Western Association of Venture Capitalists.

Oscar S. Pollock has worked in the investment business for the past twenty-five years. Since 1975 he has been a limited partner of Ingalls & Snyder, where he manages a wide range of individual and institutional accounts. He has, however, retained his original interest in investment research and follows many of his original investments directly. In his spare time he works on tax legislation for his industry. He is on the Tax Committee of The American Council for Capital Formation and the Executive Board of the Securities Industry

Association's New York District. His work on tax issues affecting individual investors has been cited in numerous publications, including the *Wall Street Journal*, the *New York Times*, and *Barrons.*

Aldona E. Robbins is senior fellow at the Institute for Political Economy. She was senior economist in the Office of Economic Policy, U.S. Department of the Treasury, where she developed a method for analyzing the effect of the social security system on the federal budget, future tax burdens, employment, and the overall performance of the U.S. economy.

Gary A. Robbins is senior fellow at the Institute for Political Economy. He was chief of the applied econometrics staff, U.S. Department of the Treasury, and the assistant to the under secretary for tax and economic affairs. He has developed new methods for analyzing the impact of taxation and government spending on economic growth.

Paul Craig Roberts is chairman of the Institute for Political Economy and holds the William E. Simon Chair in Political Economy at the Center for Strategic and International Studies, Georgetown University. During 1981–82 he was assistant secretary of the treasury for economic policy and received the Meritorious Service Award for outstanding contributions to the formulation of United States economic policy. He is credited by President Reagan with a major role in the Economic Recovery Tax Act of 1981. From 1975 to 1978 he served on the U.S. congressional staff where he drafted the Kemp-Roth bill. During 1978 through 1980 he was associate editor and columnist for the *Wall Street Journal.* He has held many academic appointments and is author of numerous articles and books including *The Supply-Side Revolution.* He is a regular columnist for *Business Week* magazine.

Thomas F. Ruhm is vice president and assistant general counsel of Bessemer Securities Corporation in New York. For the past fifteen years (following an eight-year association with a Wall Street law firm) he has been principally concerned with the legal aspects of Bessemer's venture capital and other private equity investment groups. He has spoken and written widely on the subjects of lower capital gains taxes as an incentive to long-term investment, the legal aspects of venture capital investing, and venture capital investing generally. He has been a member of the Policy Committee of the American Council for Capital Formation, Washington, D.C., since 1980 and is a frequent guest lecturer on venture capital at New York University. He is a graduate of Phillips Exeter Academy, Yale University, and Harvard Law School.

Emil M. Sunley is director of tax analysis in the National Affairs Office of Deloitte Haskins & Sells. Prior to coming to DH&S he served as deputy assistant secretary of the treasury for tax policy, 1977–1981. He serves on the Taxation Committee of the U.S. Chamber of Commerce. He served as the economic consultant to the Ad Hoc Electronics Tax Group, which recently completed its report, *High Technology Tax Policies for the 1980's.* He received a Ph.D. in economics from the University of Michigan and graduated from Amherst College.

Norman B. Ture is president of the Institute for Research on the Economics of Taxation. He was under secretary of the treasury for tax and economic affairs from March 1981 until June 1982. He came to Treasury following a long career in government, teaching, and private enterprise. Dr. Ture is the author of numerous books and articles on public policy issues, particularly in the field of tax policy. He has lectured extensively on those issues. He received his M.A. and Ph.D. degrees in economics from the University of Chicago.

Charls E. Walker is chairman of Charls E. Walker Associates, Inc., a Washington consulting firm that advises and represents some of the nation's largest enterprises. Dr. Walker served as deputy secretary of the treasury from 1969 to 1973 and as executive vice president of the American Bankers Association prior to that. Before coming to Washington, he taught finance in his home state at the University of Texas and at the Wharton School, where he gained his Ph.D. He has been an economist with the Federal Reserve System and originally served in government as principal economic adviser to President Eisenhower's Treasury Secretary Robert B. Anderson. Dr. Walker is a commentator on the PBS news show "Nightly Business Report."

James W. Wetzler is a vice president at Bear, Stearns & Co. Until 1985 he was deputy chief of staff of the Joint Committee on Taxation, U.S. Congress, where he served in various capacities since 1973. He received a Ph.D. in economics from Harvard University in 1973 and a B.S. in economics from the Wharton School, University of Pennsylvania, in 1969. He has written a number of articles on tax policy.

ABOUT THE EDITORS

Ralph Landau, former chairman of the board of The Halcon SD Group, Inc., New York City, is now with Listowel Incorporated. From 1941 to 1945 he was a process development engineer for M.W. Kellogg and head of the chemical department of Kellex Corporation, a subsidiary. In 1946 Dr. Landau became executive vice president of Scientific Design Company, Inc., of which he was cofounder and co-owner. In 1963 Scientific Design became Halcon International, Inc., with Dr. Landau as president. From 1975 he was chairman of this high-technology company in the chemicals field; it was succeeded in 1981 by The Halcon SD Group, Inc. In 1966 he cofounded the Oxirane Company with Atlantic Richfield Corporation. Oxirane has become a major petrochemical company entirely based on new technology, with seven plants producing about 5 billion pounds of products per year. In 1980 Halcon sold its half-interest in Oxirane to ARCO; Texas Eastern acquired Halcon in 1982.

Dr. Landau's numerous affiliations with universities and industry include the following: life member of the Massachusetts Institute of Technology Corporation and vice president of the National Academy of Engineering; life trustee of the University of Pennsylvania; trustee of the California Institute of Technology; trustee of the Cold Spring Harbor Laboratory; and a director of the Aluminum Company of America. He is a fellow of the faculty of the Kennedy School of Government at Harvard University and a consulting professor of economics at Stanford University. He is adjunct professor of management, technology, and society at the University of Pennsylvania.

He graduated from the University of Pennsylvania in 1937 with a bachelor of science degree and in 1941 from the Massachusetts Institute of Technology as a doctor of science; both degrees are in chemical engineering. In 1985 he was among the first recipients of the National Medal of Technology from President Ronald Reagan.

Dale W. Jorgenson is Frederic Eaton Abbe Professor of Economics at Harvard University, where he has been professor of economics since 1969. He taught at the University of California, Berkeley, from 1959 to 1969. Professor Jorgenson is a member of the National Academy of Sciences and the American Academy of Arts and Sciences. He is a fellow of the American Association for the Advancement of Science, the American Statistical Association, and the Econometric Society. Professor Jorgenson was awarded the prestigious John Bates Clark Medal of the American Economic Association in 1971 and is currently first vice president of the Econometric Society. He is author and coauthor of over 150 journal articles, contributions to books, and books in economics.